STUDENT SOLUTION MANUAL

to accompany

GENERAL STATISTICS

FOURTH EDITION

WARREN CHASE
FRED BOWN
Framingham State College

John Wiley & Sons, Inc.

New York / Chichester / Weinheim
Brisbane / Singapore / Toronto

To order books or for customer service call 1-800-CALL-WILEY (225-5945).

ISBN 0-471-28311-8

Printed in the United States of America

10 9 8 7 6 5 4 3 2 1

Printed and bound by Hamilton Printing Company

Chapter 1

1.1 A sample is a collection of some of the elements from the population, so every element in the sample is an element of the population. However, a population will contain elements that are not contained in the sample, except where the sample and the population are the same.

1.3 False   1.5   True   1.7   Inferential

1.9 Descriptive   1.11   Inferential

1.13 Descriptive   1.15   Descriptive; inferential

1.17 Statistic

1.19 (a) The collection of 10,000 voters   (b)   .35   (c)   .40
     (d) It is not surprising that the answers in parts (b) and (c) differ, as they are based on different collections of data values. The value of the statistic would most likely change if another sample were taken, because different samples give different data values. The value of the parameter would not change.

1.21 (a) A family is an element of the population, and the population size is 1000.
     (b) The sample consists of 50 families.
     (ci) $(20)(22) = 440$ (cii) $300 + 80 + 50 = 430$
     (di) $(6 + 10 + 12 + 8)/50 = .72$ (dii) $(120 + 180 + 270 + 300)/1000 = .87$

1.23 (a) Sample two elements from $\{0, 1, 2, 3\}$ and one from $\{5, 7\}$. An example is $\{1, 2, 7\}$.
     (b) Randomly select one cluster, or randomly select two clusters and sample one element from each. An example is $\{2, 3\}$.
     (c) The population size is 6. The sample size is 3 and $6/3 = 2$. If the starting point is 5, we then get $\{5, 0, 2\}$. An example is $\{5, 0, 2\}$.

1.25 (a) The population size is 100. The sample size is 10 and $100/10 = 10$. If the starting point is 00, we then get $\{00, 10, 20, 30, 40, 50, 60, 70, 80, 90\}$.
     (bi) $\{10, 37, 08, 99, 12, 66, 31, 85, 63, 73\}$
     (bii) Departments A, B, C, and D have 10, 20, 30, and 40 employees, respectively. Proportionately, select 1, 2, 3, and 4 from A, B, C, and D, respectively.
         $\{10, 37, 08, 99, 12, 66, 31, 85, 63, 44\}$
     (c) Department A is selected: $\{08, 09, 04\}$ Department B is selected: $\{10, 12, 11\}$
         Department C is selected: $\{37, 31, 44\}$ Department D is selected: $\{99, 66, 85\}$

1.27 (a) The 3–digit numbers from Appendix Table B.1 are

100	973	253	376	520	135	863	467	354	876
809	590	911	739	292	749	453	754	204	805
648	947	429	624	240	372	063	610	402	008
229	166	508	422	689	531	964	509	303	232

	Shift I	Shift II	Shift III	Totals
Male	11	9	5	25
Female	7	4	4	15
Totals	18	13	9	40

(b) The ID numbers from the males and shift I are: {072, 170, 133, 199, 291, 106, 010, 240, 148, 234, 030, 285}.

	Shift I	Shift II	Shift III	Totals
Male	12	8	4	24
Female	8	4	4	16
Totals	20	12	8	40

(c) The selected elements are: 321, 346, 371, 396, 421, 446, 471, 496, 521, 546, 571, 596, 621, 646, 671, 696, 721, 746, 771, 796, 821, 846, 871, 896, 921, 946, 971, 996, 021, 046, 071, 096, 121, 146, 171, 196, 221, 246, 271, 296. There are 16 females, and 20 are in Shift I.

(d) Shifts II and III have 300 and 200 employees, respectively. So Shift II has 3/5 of the $300 + 200 = 500$ employees. Select $(3/5)(40) = 24$ from Shift II and $(2/5)(40) = 16$ from Shift III.

1.29 (a) The new keyboard
   (b) Treatment group is those employees selected to use the new keyboards; control group is those employees selected to use the keyboards currently in use.
   (c) Yes

1.31 (a) No. Although *U.S. News* did not specify whether this was a cross sectional study or a longitudinal study, it was in all probability a cross sectional study. The heavier drinkers probably dropped out or flunked out along the way, lowering the average consumption figure for upper class students.
   (b) Select a sample of freshmen and track them through four years, recording their drinking habits. Use in your study only those who completed all four years.

1.33 Cross sectional   1.35 Longitudinal

1.37 The percentage of taxpayers in the highest bracket (50%) increased dramatically from 10% to 60%, whereas the percentage in the lowest bracket (20%) decreased significantly from 70% to 10%.

Chapter 2

2.1 (a)

x	0	1	2	3	4
f	5	5	2	2	1

(b) $(3/15)(100) = 20$

2.3 (a)

x	2	3	4	5	6	7	8	9	10	12
f	1	5	6	3	6	6	4	2	2	1

(b) $(15/36)(100) = 41.7$

2.5 (a)

Class	Cl Limits	Freq f
1	1–2	1
2	3–4	11
3	5–6	9
4	7–8	10
5	9–10	4
6	11–12	1

(b) $(5 + 6)/2 = 5.5$
(c) 4.5
(d) $10/36 = .278$

2.7

Cl Limits	Freq f
30–34	1
35–39	12
40–44	10
45–49	14
50–54	2
55-59	0
60-64	0
65-69	0
70-74	1

2.9

Cl Limits	Cl Boundaries	Cl Mark X
9.3–11.3	9.25–11.35	10.3
11.4–13.4	11.35–13.45	12.4
13.5–15.5	13.45–15.55	14.5
15.6–17.6	15.55–17.65	16.6
17.7–19.7	17.65–19.75	18.7
19.8–21.8	19.75–21.85	20.8

2.11 (a)

Class	Cl Limits	Freq f	Cl Mark X
1	4.0–5.9	2	4.95
2	6.0–7.9	5	6.95
3	8.0–9.9	10	8.95
4	10.0–11.9	11	10.95
5	12.0–13.9	0	12.95
6	14.0–15.9	1	14.95

(b) $6.0 - 4.0 = 2$
(c) $10/29 = .345$

2.13 (a)

Class	Cl Boundaries	Freq f
1	14.5–17.5	2
2	17.5–20.5	5
3	20.5–23.5	3
4	23.5–26.5	12
5	26.5–29.5	4
6	29.5–32.5	4

(b) $20.5 - 17.5 = 3$
(c) 15
(d) $5/30 = .167$

2.15 (a) 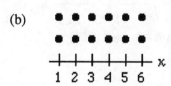 (b)

2.17 (a)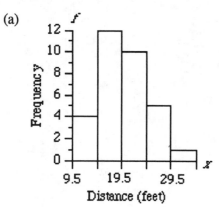

(b)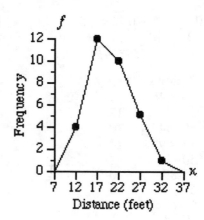

(c) Class width = 15 − 10 = 5

2.19 (a)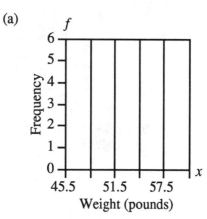

(b)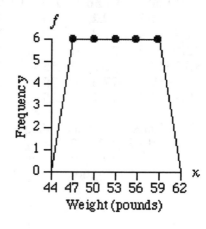

(c) Class width = 49 − 46 = 3

2.21 (a)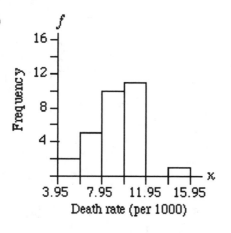

(b)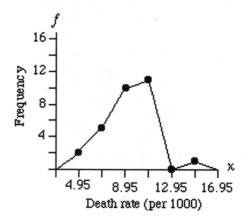

4

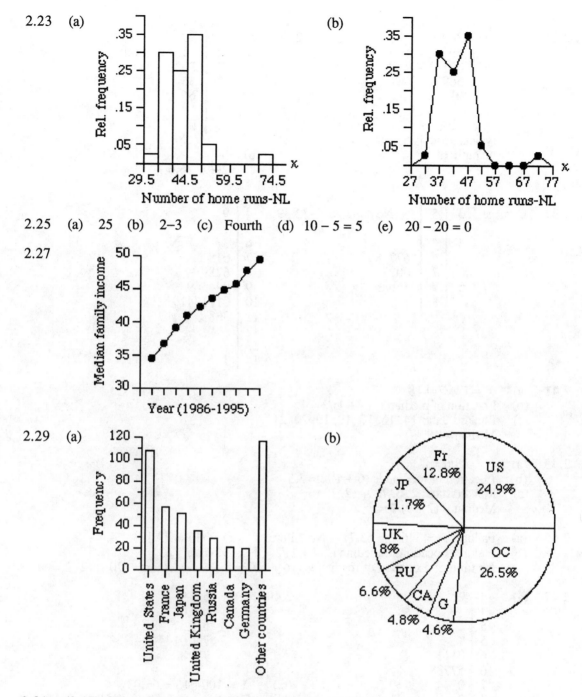

2.23 (a)

Rel. frequency / Number of home runs-NL

(b)

Rel. frequency / Number of home runs-NL

2.25 (a) 25 (b) 2–3 (c) Fourth (d) $10 - 5 = 5$ (e) $20 - 20 = 0$

2.27

Median family income / Year (1986-1995)

2.29 (a)

Frequency — United States, France, Japan, United Kingdom, Russia, Canada, Germany, Other countries

(b)

Fr 12.8% US 24.9% JP 11.7% UK 8% OC 26.5% RU 6.6% CA 4.8% G 4.6%

2.31 (2.17) Skewed to right (2.18) Bell shaped (2.19) Uniform (2.20) Skewed to left

5

2.33
```
5 | 24
5 | 59
6 | 0013334
6 | 5699
7 | 01
```

Symmetric;
typical score is 63;
no outliers

2.35 (a)
```
1 | 99
2 | 1111110000
2 | 33322
2 | 55
2 | 66
2 |
3 | 0
```

(b) 30.4
(c) 26.9; 26.6

2.37 Damage
```
        3 | 5 |          No damage
       78 | 5 |
        3 | 6 |
          | 6 | 677789
       00 | 7 | 0023
        5 | 7 | 56689
          | 8 | 1
```

2.39
```
 4 | 9
 5 | 1
 6 | 37
 7 | 018
 8 | 6789
 9 | 044569
10 | 01113479
11 | 368
12 |
13 |
14 | 3
```

2.41 (a) $\bar{x} = 126/7 = 18$
(b) Location of median is $(7 + 1)/2 = 4$
(c) Ranked data: 14, 16, 17, 19, 19, 20, 21
Median = 19

2.43 (a) $\bar{x} = 48/6 = 8$
(b) Location of median is $(6 + 1)/2 = 3.5$
(c) Ranked data: 2, 6, 7, 9, 12, 12
Median = $(7 + 9)/2 = 8$

2.45 (a) Failure: $\bar{x} = 446/7 = 63.71$ No failure: $\bar{x} = 1154/16 = 72.13$
(b) Failure: Location of median is $(7 + 1)/2 = 4$ Median = 63
No failure: Location of median is $(16 + 1)/2 = 8.5$ Median = $(70 + 72)/2 = 71$

2.47 (a)
```
4 | 2
4 |
5 |
5 | 9
6 | 0111
6 | 57788
7 | 0144
7 | 8
```

(bi) $\bar{x} = 1046/16 = 65.38$
(bii) Location of median is $(16 + 1)/2 = 8.5$
Median = $(67 + 67)/2 = 67$

The median is preferable, as the distribution is
skewed.

2.49 (a)
```
5 | 9
6 | 0111
6 | 57788
7 | 0144
7 | 8
```

(bi) $\bar{x} = 1004/15 = 66.93$
(bii) Location of median is $(15 + 1)/2 = 8$
Median = 67
The distribution is not skewed, so the mean
and the median have about the same value.
Either measure is fine.

6

2.51 (a) $\bar{x} = 14{,}912/51 = 292.39$ (b) Location of median is $(51 + 1)/2 = 26$ Median = 249
     (c) The median is preferable, as the distribution is skewed.

2.53 (a) Mean   (b) Median; the distribution is skewed.
     (c) $\bar{x} = [8(0) + 6(5) + 4(10) + 2(15) + 1(20) + 1(25) + 3(30)]/25 = 9.4$
       Location of median is $(25 + 1)/2 = 13$ Median = 5

2.55 (a) $\Sigma x = 2 + 5 + 2 + 6 + 0 = 15$
     (b) $\bar{x} = 15/5 = 3$ $\Sigma(x - \bar{x}) = (2 - 3) + (5 - 3) + (2 - 3) + (6 - 3) + (0 - 3) = 0$
       Note: In general, $\Sigma(x - \bar{x}) = 0$
     (c) $\Sigma(x - \bar{x})^2 = (2 - 3)^2 + (5 - 3)^2 + (2 - 3)^2 + (6 - 3)^2 + (0 - 3)^2 = 24$
     (d) $\Sigma x^2 = 2^2 + 5^2 + 2^2 + 6^2 + 0^2 = 69$
     (e) $(\Sigma x)^2 = (2 + 5 + 2 + 6 + 0)^2 = 15^2 = 225$
     (f) $\Sigma 7x = 7(2) + 7(5) + 7(2) + 7(6) + 7(0) = 7(2 + 5 + 2 + 6 + 0) = 7\Sigma x = 105$

2.57 About half of the families in the United States reported an income of less than \$49,687.

2.59
(a) Range = $36 - 16 = 20$

(b) $s^2 = \dfrac{6(4638) - (162)^2}{6(5)} = 52.8$

(c) $s = \sqrt{\dfrac{6(4638) - (162)^2}{6(5)}} = 7.27$

2.61
(a) Range = $23 - 10 = 13$

(b) $s^2 = \dfrac{10(2690) - (160)^2}{10(9)} = 14.44$

(c) $s = \sqrt{\dfrac{10(2690) - (160)^2}{10(9)}} = 3.80$

Note that the mean is 16, and that

$$s^2 = \frac{\Sigma(x - 16)^2}{n - 1} = \frac{130}{10 - 1} = 14.44$$

2.63
(a) Range = $140 - 116 = 24$

(b) $s^2 = \dfrac{6(102{,}356) - (782)^2}{6(5)} = 87.07$

(c) $s = \sqrt{\dfrac{6(102{,}356) - (782)^2}{6(5)}} = 9.33$

2.65 (a)

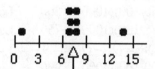

(b) Both data sets have the same mean. Data set (ii) appears to have less spread about its mean.

(ci) $s = \sqrt{\dfrac{8(636) - (60)^2}{8(7)}} = 5.15$ (cii) $s = \sqrt{\dfrac{8(536) - (60)^2}{8(7)}} = 3.51$

2.67 (a) San Diego is warmer; Bismarck has more temperature variability.

(b)
8	0	
46	1	
59	2	
	3	
37	4	
48	5	688
49	6	02389971
1	7	1

(c) Bismarck: $\bar{x} = 498/12 = 41.5$

$$s = \sqrt{\frac{12(26{,}218) - (498)^2}{12(11)}} = 22.46$$

San Diego: $\bar{x} = 762/12 = 63.5$

$$s = \sqrt{\frac{12(48{,}674) - (762)^2}{12(11)}} = 5.11$$

2.69

(a) $\sigma^2 = \dfrac{4(14-7)^2 + 4(0-7)^2}{8} = 49$

(b) $\sigma^2 = \dfrac{7(140) - (28)^2}{7^2} = 4$

(c) $\sigma^2 = \dfrac{6(94) - (20)^2}{6^2} = 4.56$

2.71 (a) $(33/100)(30) = 9.9$ Location of P_{33} is 10, so $P_{33} = 66$
(b) $(87/100)(30) = 26.1$ Location of P_{87} is 27, so $P_{87} = 92$
(c) $(25/100)(30) = 7.5$ Location of Q_1 is 8, so $Q_1 = 65$
(d) $(75/100)(30) = 22.5$ Location of Q_3 is 23, so $Q_3 = 87$
(e) $IQR = Q_3 - Q_1 = 87 - 65 = 22$
(f) $(11/30)(100) = 36.67$
(g) $(30/100)(30) = 9$ Location of D_3 is 9.5, so $D_3 = (65 + 66)/2 = 65.5$

2.73 (a) $(25/100)(51) = 12.75$ Location of Q_1 is 13, so $Q_1 = 128$
     $(75/100)(51) = 38.25$ Location of Q_3 is 39, so $Q_3 = 393$
(b) $IQR = 393 - 128 = 265$
(c) $(50/100)(51) = 25.5$ Location of Q_2 is 26, so $Q_2 = 249$
(d) $(95/100)(51) = 48.45$ Location of P_{95} is 49, so $P_{95} = 697$
(e) $(10/51)(100) = 19.61$

2.75 (a) $(25/100)(1000) = 250$ Location of Q_1 is 250.5, so $Q_1 = (4.3 + 4.3)/2 = 4.3$
     $(50/100)(1000) = 500$ Location of Q_2 is 500.5, so $Q_2 = (4.6 + 4.6)/2 = 4.6$
     $(75/100)(1000) = 750$ Location of Q_3 is 750.5, so $Q_3 = (4.9 + 4.9)/2 = 4.9$
(b) $IQR = Q_3 - Q_1 = 4.9 - 4.3 = .6$
(c) $(90/100)(1000) = 900$ Location of P_{90} is 900.5, so $P_{90} = (5.2 + 5.2)/2 = 5.2$
(d) $(985)(100)/1000 = 98.5$

2.77 (a) Location of Q_1 is $(25/100)(40 + 1) = 10.25$ $Q_1 = 39 + (.25)(39 - 39) = 39$
     Location of Q_3 is $(75/100)(40 + 1) = 30.75$ $Q_3 = 48 + (.75)(49 - 48) = 48.75$
(b) $IQR = 48.75 - 39 = 9.75$
(c) The mean and the median have about the same value indicating that the data are not strongly skewed.

2.79 $L = 33$ $Q_1 = 65$; $(50/100)(30) = 15$ Location of Q_2 is 15.5
$Q_2 = (78 + 81)/2 = 79.5$ $Q_3 = 87$ $H = 95$ Q_1 and Q_3 were found in Exercise 2.71

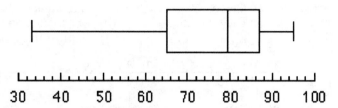

The distribution of data is skewed to the left.

2.81 $L = 4.9$ $(25/100)(29) = 7.25$ Location of Q_1 is 8 $Q_1 = 8.6$
$(50/100)(29) = 14.5$ Location of Q_2 is 15 $Q_2 = 9.5$
$(75/100)(29) = 21.75$ Location of Q_3 is 22 $Q_3 = 10.3$ $H = 14.3$

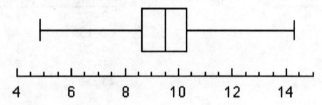

The distribution of data is nearly symmetric.

2.83 $L = 24$ $(25/100)(51) = 12.75$ Location of Q_1 is 13 $Q_1 = 128$
$(50/100)(51) = 25.5$ Location of Q_2 is 26 $Q_2 = 249$
$(75/100)(51) = 38.25$ Location of Q_3 is 39 $Q_3 = 393$
$H = 1160$ IQR $= 393 - 128 = 265$ $1.5 \times$ IQR $= 397.5$
$Q_1 - 1.5 \times$ IQR $= -269.5$ $Q_3 + 1.5 \times$ IQR $= 790.5$ Suspected outliers are 955 and 1160

(a)

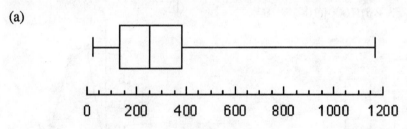

(b)

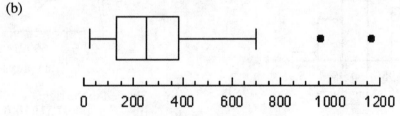

(c)   The distribution of data is skewed to the right.

2.85 (a) San Diego has a larger median
(b) San Diego
(c) Bismarck: 20.5 San Diego: 59
(d) Bismarck: 61 San Diego: 68.5

2.87 (a) Yes; the July-December boxplot is shifted to the left compared with the January-June boxplot.
(b) Yes; The two boxplots are quite similar in shape and location.

2.89 (a)

x	4.1	4.3	4.4	4.5	4.6	4.7	4.8	4.9	5.1	5.5	5.7
f	1	3	1	3	2	4	1	1	2	1	1

(b) $(16/20)(100) = 80$ (c)

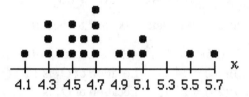

2.91

x	0	1	2	3	4	5
f	2	2	6	6	2	2

2.93

Cl Limits	Freq f
0–6	3
7–13	4
14–20	4
21–27	5
28–34	4

2.95 (a)

Cl Limits	Freq f
4.0–4.4	377
4.5–4.9	425
5.0–5.4	160
5.5–5.9	33
6.0–6.4	5

(b) $4.5 - 4.0 = .5$
(c) $[(160 + 33 + 5)/1000](100) = 19.8$

(d)

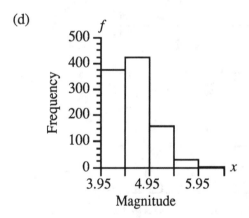

(e) Skewed to the right

2.97 (a)

Cl Limits	Freq f
0–199	21
200–399	19
400–599	6
600–799	3
800–999	1
1000–1199	1

(b)

(c) Skewed to the right
(d) $[(6 + 3 + 1 + 1)/51](100) = 21.6$

2.99 (a)

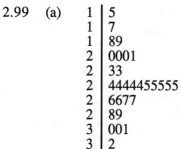

```
1 | 5
1 | 7
1 | 89
2 | 0001
2 | 33
2 | 4444455555
2 | 6677
2 | 89
3 | 001
3 | 2
```

  (b) Symmetric

2.101

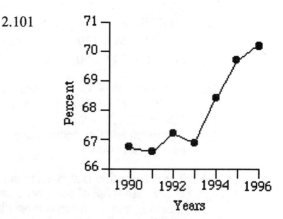

2.103 (a)

  (b)

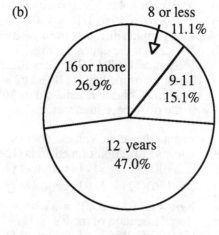

2.105 (a) 7   (b) $65.5 - 63.5 = 2$ (c) 42   (d) 66–67   (e) $9 - 3 = 6$
  (f) Draw a horizontal line from .50 on the Cumulative Relative Frequency axis to the graph; then a vertical line from this point on the graph to the Heights axis. The value on the Heights axis is about 69.
  (g) Draw a vertical line from 71.5 on the Heights axis to the graph; then a horizontal line from this point on the graph to the Cumulative Relative Frequency axis. The point of intersection on the Cumulative Relative Frequency axis is approximately .83, or 83%.

2.107 (a) $\bar{x} = 4614/65 = 70.98$ (b) Location of median is $(65 + 1)/2 = 33$ Median = 71
  (c) $s^2 = \dfrac{65(327,928) - (4614)^2}{65(64)} = 6.33$ (d) $s = \sqrt{\dfrac{65(327,928) - (4614)^2}{65(64)}} = 2.52$
  (e) $(25/100)(65) = 16.25$ Location of Q_1 is 17, so $Q_1 = 69$
       $(75/100)(65) = 48.75$ Location of Q_3 is 49, so $Q_3 = 72$
  (f) IQR $= 72 - 69 = 3$ (g) $(90/100)(65) = 58.5$ Location of P_{90} is 59 $P_{90} = 75$
  (h) $(3)(100)/65 = 4.6$

2.109 (a) $(25/100)144 = 36$ Location of Q_1 is 36.5, so $Q_1 = \dfrac{727 + 727}{2} = 727$

       $(75/100)144 = 108$ Location of Q_3 is 108.5, so $Q_3 = \dfrac{1124 + 1124}{2} = 1124$

11

(b) IQR = 1124 − 727 = 397
(c) (10/100)144 = 14.4   Location of P_{10} is 15, so P_{10} = 545
     (90/100)144 = 129.6 Location of P_{90} is 130, so P_{90} = 1289
(di) (72)(100)/144 = 50   (dii) (98)(100)/144 = 68.1

2.111 (a) $\bar{x}$ = 1727/40 = 43.17   Location of median is (40 + 1)/2 = 20.5
           Median = (43 + 44)/2 = 43.5   MIN = L = 31, and MAX = H = 70
      (b) IQR = 47 − 38 = 9
      (c) The mean and the median have about the same value indicating that the data are not strongly skewed.

2.113 (a) The distribution is skewed to the right with the populations of China and India much larger than those of the other countries. The median would appear to be a better measure, and the mean is most likely larger than the median.
      (b) $\bar{x}$ = $\Sigma x/n$ = 4175.3/20 = 208.765 million
      (c) $\bar{x}$ = $\Sigma x/n$ = 1986.1/18 = 110.34 million
           This is much different from the mean population in part (b).
      (d) Notice that the data are ranked. Location of median is (20 + 1)/2 = 10.5
           Median = (107.1+ 97.6)/2 = 102.35 million
      (e) Location of median is (18 + 1)/2 = 9.5 Median = (97.6 + 84.1)/2 = 90.85 million
           Deleting China and India did make a difference in medians, but not nearly as much as in the difference in means.

2.115 We compute the actual values. Your estimates should be fairly close.
      (ai) Albany: Location of median is (12+1)/2 = 6.5   Median = (46.6 + 50.5)/2 = 48.55
            (25/100)(12) = 3   Location of Q_1 is 3.5 Q_1 = (26.5 + 33.6)/2 = 30.05
            (75/100)(12) = 9   Location of Q_3 is 9.5 Q_3 = (61.2 + 66.7)/2 = 63.95
            IQR = 63.95 − 30.05 = 33.90
            Reno: Location of median is (12+1)/2 = 6.5   Median = (46.4 + 50.3)/2 = 48.35
            (25/100)(12) = 3   Location of Q_1 is 3.5 Q_1 = (37.4 + 39.7)/2 = 38.55
            (75/100)(12) = 9   Location of Q_3 is 9.5 Q_3 = (60.2 + 62.4)/2 = 61.30
            IQR = 61.30 − 38.55 = 22.75
      (aii) Reno; the medians are the same but the variability in Reno temperatures is smaller.
      (b)   Yes; the mean temperatures are about the same and the variability in Reno temperatures is smaller.

2.117 (a) Chicago in both cases
      (b) In both cases, the location of the median is (12 + 1)/2 = 6.5.
           Boston:
           Ranked data: 10.7, 10.9, 11.3, 11.4, 12.0, 12.2, 12.9, 13.3, 13.7, 13.9, 14.1, 14.2
           Median = (12.2 + 12.9)/2 = 12.55
           Chicago:
           Ranked data: 8.1, 8.1, 8.7, 9.1, 9.8, 10.6, 10.9, 10.9, 11.5, 11.6, 11.9, 12.1
           Median = (10.6 + 10.9)/2 = 10.75
      (c) Boston: $\bar{x}$ = 150.6/12 = 12.55   Chicago: $\bar{x}$ = 123.3/12 = 10.28

2.119 (a) Median height increases as pitch decreases   (b)   75%   (c)   No

2.121 $\bar{x}$ = 43.17 and s = 6.75

	Within	No. of data	%	Empirical Rule %
1 S.D.	(36.42 − 49.92)	32	80	68
2 S.D.'s	(29.67 − 56.67)	39	97.5	95
3 S.D.'s	(22.92 − 63.42)	39	97.5	99.7

2.123 (a) $400 - 3(25) = 325$ and $400 + 3(25) = 475$ So $k = 3$ and $(1 - (1/3^2))(1000) = 888.9$
By Chebyshev's theorem, at least 889 data values lie between 325 and 475.

(b) $400 - 4(25) = 300$ and $400 + 4(25) = 500$ So $k = 4$ and $(1 - (1/4^2))(1000) = 937.5$
By Chebyshev's theorem, at least 938 data values lie between 300 and 500.

(c) $400 - 2(25) = 350$ and $400 + 2(25) = 450$ So $k = 2$ and $(1 - (1/2^2))(1000) = 750$
By Chebyshev's theorem, at least 750 data values lie between 350 and 450. At most
250 data values are smaller than 350 or larger than 450.

Chapter 3

3.1   (a) $b_0 = 4$ and $b_1 = 7$ (b) $b_0 = 4$ and $b_1 = 3$, as $y = 4 + 3x$

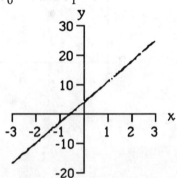
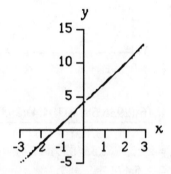

(c) $b_0 = 9/2$ and $b_1 = -2$, as $y = 9/2 - 2x$ (d) $b_0 = 13$ and $b_1 = 5$, as $y = 13 + 5x$

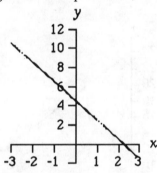
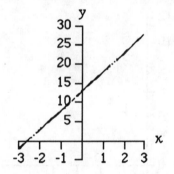

(e) $b_0 = 8$ and $b_1 = 0$, as $y = 8 + 0x$

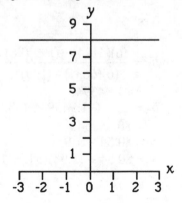

13

3.3 (b) $b_1 = \dfrac{(6)(96) - (24)(30)}{(6)(112) - (24)^2} = -1.5$

$b_0 = 5 - (-1.5)(4) = 11$

$\hat{y} = 11 - 1.5x$

(a,c)

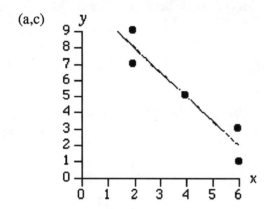

3.5 (b) $b_1 = \dfrac{(4)(160) - (12)(40)}{(4)(46) - (12)^2} = 4$

$b_0 = 10 - (4)(3) = -2$

$\hat{y} = -2 + 4x$

(a,c)

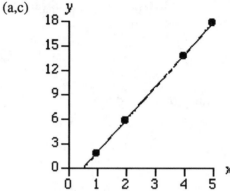

3.7

$b_1 = \dfrac{(6)(2936.68) - (104.4)(153)}{(6)(1933.120) - (104.4)^2} = 2.354839$

$b_0 = 25.5 - (2.354839)(17.4) = -15.474$

$\hat{y} = -15.474 + 2.355x$

3.9

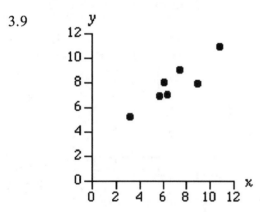

$b_1 = \dfrac{(7)(410.14) - (49.5)(54.6)}{(7)(386.790) - (49.5)^2}$

$= .654073$

$b_0 = 7.8 - (.654073)(7.071429)$

$= 3.174769$

$\hat{y} = 3.175 + .654x$

3.11

(a)

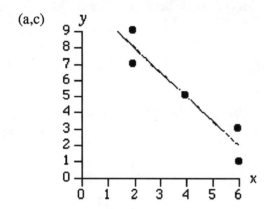

Yes

(b) $b_1 = \dfrac{(6)(3124) - (199)(96)}{(6)(6631) - (199)^2}$

$= -1.945946$

$b_0 = 16 - (-1.945946)(33.1666667)$

$= 80.540543$

$\hat{y} = 80.54 - 1.95x$

(c) $\hat{y} = 80.54 - (1.95)(34) = 14.24$

14

3.13

(a)

Yes

(b) $b_1 = \dfrac{(8)(19,425.13) - (370.6)(412.2)}{(8)(17,467.68) - (370.6)^2}$

$= 1.101223$

$b_0 = 51.525 - (1.101223)(46.325)$

$= .510845$

(c) $\hat{y} = .511 + 1.101x$

$(1.101)(\$1000) = \$1101$

3.15 (a)

(b) $b_1 = \dfrac{(23)(446) - (1600)(7)}{(23)(112,400) - (1600)^2} = -.037381$

$b_0 = .304348 - (-.037381)(69.565217)$

$= 2.904765$

(c) $\hat{y} = 2.90 - (.0374)(53) \doteq .92$

3.17

(a) $\hat{y} = 48.7075 - 8.36459x$

(b) $\hat{y} = 48.7075 - (8.36459)(3)$

$= 23.6$ MPG

3.19 (a) Yes

(b) $\hat{y} = 283.61 + 2.44x$

(c) 2.44 inches

3.21 (a) Yes

(b) $\hat{y} = -1.2329 + .0070583x$

(c) $\hat{y} = -1.2329 + (.0070583)(500) = 2.30$

3.23 (a) Yes

(b) $b_1 = \dfrac{(11)(18,267,023) - (56,562)(2611)}{(11)(456,525,234) - (56,562)^2} = .029220$

$b_0 = \dfrac{2611}{11} - (.029220)\left(\dfrac{56,562}{11}\right) = 87.114396$

$\hat{y} = 87.114 + .029x$

(c) $\hat{y} = 87.114 + (.029)(3500) = 189$

(d) Now remove the observation $x = 14235$, $y = 425$

$b_1 = \dfrac{(10)(12,217,148) - (42,327)(2186)}{(10)(253,890,009) - (42,327)^2} = .039668$

$b_0 = \dfrac{2186}{10} - (.039668)\left(\dfrac{42,327}{10}\right) = 50.697256$

$\hat{y} = 50.697 + (.040)(3500) = 191$

The estimates 189 and 191 are so close that either estimate will do.

3.25 (a) $r = \dfrac{(6)(96) - (24)(30)}{\sqrt{(6)(112) - (24)^2}\sqrt{(6)(190) - (30)^2}} = -.948683$ (b) $r^2 = (-.948683)^2 \doteq .90$

About 90% of the variation in y is explained by the linear relationship with x.

3.27 (a) $r = \dfrac{(7)(54) - (7)(42)}{\sqrt{(7)(13) - (7)^2}\sqrt{(7)(286) - (42)^2}} = .840168$ (b) $r^2 = (.840168)^2 \doteq .71$

About 71% of the variation in y is explained by the linear relationship with x.

3.29 (a) $r = \dfrac{(4)(160) - (12)(40)}{\sqrt{(4)(46) - (12)^2}\sqrt{(4)(560) - (40)^2}} = 1$ (b) $r^2 = 1^2 = 1$

The total variation (100%) in y is explained by the linear relationship with x.

3.31 (a) $r = \dfrac{(9)(7439.37) - (41.56)(1416.1)}{\sqrt{(9)(289.4222) - (41.56)^2}\sqrt{(9)(232,498.97) - (1416.1)^2}} = .926345$

(b) $r^2 = (.926345)^2 \doteq .858$ About 85.8% of the variation in the incidence of cancer is explained by the linear relationship with the index of exposure.

3.33 $r = \dfrac{(5)(403) - (15)(174)}{\sqrt{(5)(55) - (15)^2}\sqrt{(5)(7754) - (174)^2}} = -.913011$

The coefficient of determination $r^2 = (-.913011)^2 \doteq .83$. About 83% of the variation in percentage of residents infected is explained by the linear relationship with the distance people live from the reservation.

3.35 (a) $r = \dfrac{(11)(18,267,023) - (56,562)(2611)}{\sqrt{(11)(456,525,234) - (56,562)^2}\sqrt{(11)(818,149) - (2611)^2}} = .844416$

(b) $r^2 = (.844416)^2 \doteq .71$ About 71% of the variation in the number of openings is explained by the linear relationship with the number of full-and part-time employees.

3.37 (a) $r = \dfrac{(24)(37,939.3) - (13,557)(66.1)}{\sqrt{(24)(7,743,167) - (13,557)^2}\sqrt{(24)(187.93) - (66.1)^2}} = .849449$

(b) $r^2 = (.849449)^2 \doteq .722$ About 72.2% of the variation in GPA is explained by the linear relationship with Verbal SAT score.

3.39 (a) There appears to be a negative linear relationship. Yes
     (b) Approximately 81.4% of the variation in PRO is explained by the linear relationship with UNEMP.
     (c) Notice that the scatter diagram indicates a negative linear relationship between the variables.
     $r = -\sqrt{.814} = -.902$

3.41
(a)

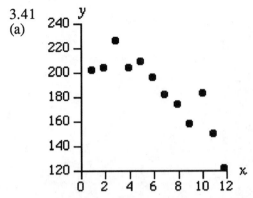

It appears that people born in the latter part of the year were more likely to be called sooner. The process is not random.

(b)

$$r = \frac{(12)(13,306.5) - (78)(2202.4)}{\sqrt{(12)(650) - (78)^2}\sqrt{(12)(413,700.24) - (2202.4)^2}}$$
$$= -.866393$$

$r^2 = (-.866393)^2 = .751$

About 75.1% of the variation in means of the ranks for the days is explained by the linear relationship with month.

3.43
(a)

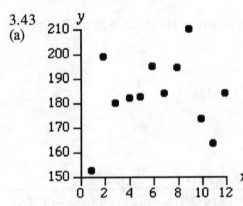

The process appears to be random.

(b)
$$r =$$
$$\frac{(12)(14,350.5) - (78)(2196.7)}{\sqrt{(12)(650) - (78)^2}\sqrt{(12)(404,847.29) - (2196.7)^2}}$$
$$= .115302$$

$r^2 = (.115302)^2 \doteq .013$

About 1.3% of the variation in means of the ranks for the days is explained by the linear relationship with month.

3.45 (a) SSR = TSS − SSE = 50 − 20 = 30 r^2 = SSR/TSS = 30/50 = .60

   (b) SSR = (TSS)(r^2) = 40(.2) = 8   SSE = TSS − SSR = 40 − 8 = 32

   (c) TSS = SSR/r^2 = 25/1 = 25   SSE = TSS − SSR = 25 − 25 = 0

   (d) TSS = SSR + SSE = 30 + 0 = 30 r^2 = SSR/TSS = 30/30 = 1

3.47 (a) $b_1 = \dfrac{(5)(85) - (25)(20)}{(5)(143) - (25)^2} = -.833333$ $b_0 = 4 - (-.833333)(5) = 8.166665$

   $\hat{y} = 8.17 - .83x$

   (b) $r = \dfrac{(5)(85) - (25)(20)}{\sqrt{(5)(143) - (25)^2}\sqrt{(5)(96) - (20)^2}} = -.883883$ $r^2 = (.883883)^2 = .781$

   About 78% of the variation in y is explained by the linear relationship with x.

3.49 (a)

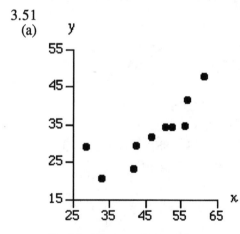

Positive linear relationship

(b) $r = \dfrac{(7)(7901) - (174)(218)}{\sqrt{(7)(6136) - (174)^2}\sqrt{(7)(11,056) - (218)^2}} = .892957$

$r^2 = (.892957)^2 = .797$

About 80% of the variation in male population death rate is explained by the linear relationship with average cholesterol intake.

(c) $b_1 = \dfrac{(7)(7901) - (174)(218)}{(7)(6136) - (174)^2} = 1.370701$

$b_0 = 31.142857 - (1.370701)(24.857143) = -2.928854$

$\hat{y} = -2.93 + 1.37x$

(d) $\hat{y} = -2.93 + (1.37)(40) = 51.87$

3.51
(a)

y

Positive linear relationship

(b) $r =$

$\dfrac{(10)(15,985.69) - (473.1)(324.4)}{\sqrt{(10)(23,389.21) - (473.1)^2}\sqrt{(10)(11,098.78) - (324.4)^2}}$
$= .838753$

$r^2 = (.838753)^2 \doteq .704$

About 70.4% of the variation in median household income is explained by the linear relationship with percentage of residents who are college graduates.

(c)

$b_1 = \dfrac{(10)(15,985.69) - (473.1)(324.4)}{(10)(23,389.21) - (473.1)^2} = .633984$

$b_0 = 32.44 - (.633984)(47.31) = 2.446217$

$\hat{y} = 2.446 + .634x$

(d)

$\hat{y} = 2.446 + (.634)(40) = 27.806$, that is

$(27.806)(\$1000) = \$27,806$

3.53 (a) $r^2 = .15201$ About 15.2% of the variation in number of home runs is explained by the number of hits.

(b) $\hat{y} = -47.493453 + .148406x$ $\hat{y} = -47.493453 + (.148406)(1500) = 175$

(c) Notice that the slope of the line is approximately .15. This means that each unit increase in x (each hit) increases y (the number of home runs) by .15. So a change of 7 additional hits would increase the number of home runs by $(7)(.15) = 1.05$, or about 1, on average.

3.55 (a) $r^2 = .61195$
About 61.2% of the variation in number of home runs is explained by the number of hits.

(b) $b_1 = .401515$ is positive and has the same sign as r
$r = \sqrt{.61195} = .782$

3.57 (a) $r = \dfrac{(203)(2243.7) - (169.1)(2240.4)}{\sqrt{(203)(173.6) - (169.1)^2}\,\sqrt{(203)(30,390) - (2240.4)^2}} = .876501$

$r^2 = (.876501)^2 = .768$
Approximately 76.8% of the variation in carbon monoxide content is explained by the linear relationship with nicotine content.

(b) $b_1 = \dfrac{(203)(2243.7) - (169.1)(2240.4)}{(203)(173.6) - (169.1)^2} = 11.528675$

$b_0 = 11.036453 - (11.528675)(.833005) = 1.433009$
$\hat{y} = 1.433 + 11.529x$

(c) $\hat{y} = 1.433 + (11.529)(1) = 12.962$

3.59 (a) $s_x^2 = \dfrac{(5)(63) - (15)^2}{(5)(4)} = 4.5 \qquad s_x = \sqrt{4.5} = 2.121320$

$s_y^2 = \dfrac{(5)(114) - (20)^2}{(5)(4)} = 8.5 \qquad s_y = \sqrt{8.5} = 2.915476$

(b) $r = \dfrac{(5)(77) - (15)(20)}{\sqrt{(5)(63) - (15)^2}\,\sqrt{(5)(114) - (20)^2}} = .687184$

(c) $b_1 = \dfrac{(5)(77) - (15)(20)}{(5)(63) - (15)^2} = .944444 \qquad\qquad$ Now, $\bar{x} = 15/5 = 3$, and $\bar{y} =$
$20/5 = 4$. Therefore, $b_0 = 4 - (.944444)(3) = 1.166668$. So $\hat{y} = 1.17 + .94x$.

3.61 (a) $\hat{y} = 61 + (.6)(5/5)(x - 60) = 61 + (.6)(x - 60)$
(b) $\hat{y} = 61 + (.6)(80 - 60) = 73$
(c) $\hat{y} = 61 + (.6)(40 - 60) = 49$

Chapter 4

4.1 (a) $S = \{O, A, B, AB\}$
(b) $S = \{OO, OA, OB, O\underline{AB}, AO, AA, AB, A\underline{AB}, BO, BA, BB, B\underline{AB}, \underline{AB}O, \underline{AB}A,$
$\underline{AB}B, \underline{AB}\,\underline{AB}\}$

4.3 (a) $E = \{4260, 4261, 4262, 4263, 4264, 4265, 4266, 4267, 4268, 4269\}$
(b) $E = \{2468, 2648, 4268, 4628, 6248, 6428\}$

4.5 $S = \{xX, xY, XX, XY\}$

4.7 $S = \{0A, 0B, 0C, 0D, 0E, 1A, 1B, 1C, 1D, 1E\}$
(a) $E = \{0D, 0E\}$ (b) $E = \{1A, 1B, 1C, 1D\}$
(c) $E = \{0A, 0B, 1A, 1B\}$ (d) $E = \{1A, 1B, 1C, 1D, 1E\}$

4.9 (a) $E = \{M1, M2, M3, F1, F2, F3\}$ (b) $E = \{F1, F2, F3, F4, F5, F6, M4, M5, M6\}$
(c) $E = \{M6\}$ (d) $E = \{M3\}$

4.11 Recall that $n(S) = 36$
(a) Let E be the event that the sum is less than 5.
$E = \{(1,1), (1,2), (1,3), (2,1), (2,2), (3,1)\}$ $P(E) = 6/36 = 1/6$
(b) Let E be the event of one or two 6's
$E = \{(1,6), (2,6), (3,6), (4,6), (5,6), (6,1), (6,2), (6,3), (6,4), (6,5), (6,6)\}$
$P(E) = 11/36$
(c) Let E be the event neither is a 6 $E = \{(1,1), (1,2), (1,3), (1,4), (1,5), (2,1), (2,2),$
$(2,3), (2,4), (2,5), (3,1), (3,2), (3,3), (3,4), (3,5), (4,1), (4,2), (4,3), (4,4), (4,5),$
$(5,1), (5,2), (5,3), (5,4), (5,5)\}$ $P(E) = 25/36$
(d) Let E be the event that the sum is either 7 or 11
$E = \{(1,6), (2,5), (3,4), (4,3), (5,2), (6,1), (5,6), (6,5)\}$ $P(E) = 8/36 = 2/9$

4.13 (a) $S = \{0, 1, 2, 3, 4, 5, 6, 7, 8, 9\}$
(b) $E = \{1, 3, 5, 7, 9\}$ $P(E) = 5/10$
(c) $E = \{7, 8, 9\}$ $P(E) = 3/10$

4.15 Let b = boy and g = girl. This Exercise is much like Exercise 4.14, with $b = E$ and
$g = O$. $S = \{bbbb, bbbg, bbgb, bgbb, gbbb, bbgg, bgbg, bggb, gbbg, gbgb, ggbb, gggb,$
$ggbg, gbgg, bggg, gggg\}$, where the list for a particular family is from youngest to oldest.
Now $n(S) = 16$.
(a) Let E be the event of two boys. $E = \{bbgg, bgbg, bggb, gbbg, gbgb, ggbb\}$
$P(E) = 6/16$
(b) Let E be the event the oldest two children are boys
$E = \{bbbb, bgbb, gbbb, ggbb\}$ $P(E) = 4/16$
(c) Let E be the event all are girls. $E = \{gggg\}$ $P(E) = 1/16$
(d) Not all are girls is the same as one or more boys. Let E be the event of one or more
boys. $P(E) = 15/16$

4.17 $n(S) = 395 + 75 + 833 + 209 + 999 + 298 = 2809$
(a) Let E be the event of receiving a master's degree.
$P(E) = (75 + 209 + 298)/2809 = 582/2809 = .2072$
(b) Let E be the event of receiving a degree in 1970.
$P(E) = (833 + 209)/2809 = 1042/2809 = .3710$

4.19 $(26)^3(10)^3 = 17,576,000$

4.21 (a) $9(10)^2(9)(10)^6 = 9^2(10)^8 = 8,100,000,000$ (b) $1^3(9)(10)^6 = 9(10)^6 = 9,000,000$
(c) $10^4 = 10,000$

4.23 (a) $\dfrac{1^5}{6^3(2)^2} = .0012$ (b) $\dfrac{1^3(2)^2}{6^3(2)^2} = .0046$ (c) $\dfrac{6^3(1)^2}{6^3(2)^2} = .25$

4.25 (a) A, B is a mutually exclusive pair, as the selected card can't be an ace and a face card.
Notice that the ace of clubs belongs to both A, C, while the jack of clubs belongs to
both B, C.
(b) $P(A \text{ or } B) = (4/52) + (12/52) = 16/52$

20

4.27   The total number (in thousands) is $n(S) = 53{,}312 + 35{,}245 + 60{,}554 + 36{,}573 = 185{,}684$
      (a) $P(B) = (53{,}312 + 35{,}245)/185{,}684 = .477$
      (b) $P(\overline{A}) = (35{,}245 + 36{,}573)/185{,}684 = .387$
      (c) $P(\overline{A} \text{ and } \overline{B}) = 36{,}573/185{,}684 = .197$

4.29   (ai) $P(A \text{ and } B) = P(A)P(B) = (4/52)(4/52) = .0059$
      (aii) The events B, C are mutually exclusive.
            $P(B \text{ or } C) = P(B) + P(C) = 4/52 + 4/52 = .1538$
      (b)   The events A, B, D are independent.
            $P(A \text{ and } B \text{ and } D) = P(A)P(B)\, P(D) = (4/52)(4/52)(48/52) = .0055$

4.31   Let event A = at least one child will have the disease. So $\overline{A}$ is the event that no children
      will have the disease, and $P(\overline{A}) = 9/10.$ $P(A) = 1 - (9/10)^4 = .3439$

4.33 $P(L \text{ and } M \text{ and } N) = P(L)P(M)\, P(N) = (.337)(.796)(.016) = .0043$
      Note that this is fewer than $1/200 = .005$

4.35   Let events A = male, and B = under the age of 25
      $P(A \text{ or } B) = P(A) + P(B) - P(A \text{ and } B) = (1500/4000) + (3600/4000) - (1200/4000)$
      $= .9750$

4.37   (a) $P(A \text{ and } B) = P(A)P(B \mid A) = (4/52)(3/51) = .0045$
      (b) $P(B) = (4/52)(3/51) + (48/52)(4/51) = .0769.$
            No. $P(B) = .0769 \neq P(B \mid A) = 3/51 = .0588$
      (c) $P(A \text{ or } B) = P(A) + P(B) - P(A \text{ and } B) = 4/52 + 4/52 - (4/52)(3/51) = .1493$

4.39   Let events A = colorblind, and B = male. The sample space $S = \{xX, xY, XX, XY\}$.
      A male and female will be colorblind if they possess xY and xx, respectively.
      (a)   Given a male, the sample space is $\{xY, XY\}$. The event of colorblindness given a
            male is $\{xY\}$. So $P(\text{colorblind given a male}) = 1/2$.
      (b)   Given a female, the sample space is $\{xX, XX\}$. As the sample space does not contain
            the pair xx, $P(\text{colorblind given a female}) = 0$.

4.41   Let events A = male, and B = employed more than 20 years.
      $P(A \mid B) = P(A \text{ and } B)/P(B) = P(B \mid A)P(A)/P(B) = (30/600)(600/1000)/(70/1000) = .4286$
      We could have looked at this in the following way:

	Employed ≤ 20 yrs	Employed > 20 yrs	Total
Males	570	30	600
Females	360	40	400
Total	930	70	1000

      $P(A \mid B) = 30/(30 + 40) = .4286$

4.43   (a) $P(\text{female} \mid \text{no opinion}) = 10/(6 + 10) = .6250$
      (b) $P(\text{approves} \mid \text{male}) = 21/(21 + 6 + 12) = .5385$
      (c) $P(\text{male} \mid \text{did not approve}) = 12/(12 + 7) = .6316$
      (d) $P(\text{male and approves}) = P(\text{male})P(\text{approves} \mid \text{male})$
            $= [(21 + 6 + 12)/(21 + 6 + 12 + 14 + 10 + 7)](21/39) = .3000$

4.45 Let events A = husband earns less than \$30,000, and B = wife earns less than \$30,000.
    (a) $P(A \mid B) = 430/(430 + 410) = .5119$
    (b) $P(\overline{B} \mid \overline{A}) = 100/(410 + 100) = .1961$
    (c)   No, because $P(A) = (430 + 60)/1000 = .49 \neq P(A \mid B) = .5119$

4.47 Let b = boy and g = girl. Let $S = \{bb, bg, gb, gg\}$ with each pair representing oldest and youngest. For example, bg means the oldest is a boy and the youngest a girl.
    (a)   View the sample space as $\{bb, bg\}$. $P[(A \text{ and } B) \mid A] = 1/2$
    (b)   View the sample space as $\{bb, bg, gb\}$. $P[(A \text{ and } B) \mid (A \text{ or } B)] = 1/3$

4.49 (a) $(6)(5)(4)(3)(2)(1) = 720$ (b) $12!/(12 - 3)! = 12!/9! = (12)(11)(10) = 1320$
    (c) $5!/(5 - 5)! = 5!/0! = 5!/1 = (5)(4)(3)(2)(1) = 120$
    (d) $12!/(3!)(12 - 3)! = 12!/(3!)(9!) = (12)(11)(10)/(3)(2)(1) = 220$
    (e) $12!/(9!)(12 - 9)! = 12!/(9!)(3!) = (12)(11)(10)/(3)(2)(1) = 220$
    (f) $5!/(5!)(5 - 5)! = 5!/(5!)(0!) = 5!/(5!)(1) = 1$

4.51 $10! = 3,628,800$

4.53 (a) ${}_4C_3/{}_{10}C_3 = 4/120 = .0333$ (b) $({}_6C_2)({}_4C_1)/{}_{10}C_3 = (15)(4)/120 = 1/2$

4.55 (a) $1/6! = .0014$ (b) $(1)(1)(1)(3!)/6! = .0083$
    (c)   Let E be the event that A, B appear next to each other. $P(E) = (2!)(5)(4!)/6! = 1/3$

4.57 (a) $({}_4C_4)({}_{48}C_9)/{}_{52}C_{13} = \dfrac{(13)(12)(11)(10)}{(52)(51)(50)(49)} = .0026$, or 26 in 10,000
    (b) $({}_{13}C_7)({}_{13}C_6)({}_{26}C_0)/{}_{52}C_{13} = .000005$, or 5 in 1 million

4.59 $S = \{11, 12, 13, 14, 21, 22, 23, 24, 31, 32, 33, 34, 41, 42, 43, 44\}$

4.61 Let events W = win, and L = lost.   (a) $S = \{WW, WL, LW, LL\}$
    (bi) $\{WW\}$ (bii) $\{WL, LW\}$

4.63 (a) $P(\text{your birth date}) = 1/365$ (b) $P(\text{day in July}) = 31/365$
    (c) $P(\text{first day of month}) = 12/365$
    (d)   Consider the months January, June, and July.
          $P(\text{day in month beginning with J}) = (31 + 30 + 31)/365 = 92/365$

4.65 $n(S) = 3,838,380 + 3,948,048 + 1,370,850 + 197,600 + 11,700 + 240 + 1 = 9,366,819$
    (a)   Let E be the event of matching 3 numbers. $P(E) = 197,600/9,366,819 = .0211$
    (b)   Let E be the event of matching 0 or 1 numbers.
          $P(E) = (3,838,380 + 3,948,048)/9,366,819 = .8313$
    (c)   Let E be the event of matching either 4, 5, or 6 numbers.
          $P(E) = (11,700 + 240 + 1)/9,366,819 = .0013$, which is about 13 times in 10,000

4.67 Recall that the sample space $S = \{xX, xY, XX, XY\}$. Let events A = youngest is colorblind, and B = oldest is colorblind. Assume A, B are independent events. As xY is the only outcome that yields colorblindness, the probability of being colorblind is 1/4.
    (a) $P(\overline{A} \text{ and } B) = (3/4)(1/4) = .1875$
    (b)   Use the result in part (a). $P(\text{exactly one is colorblind}) = 2(3/16) = .3750$
    (c) $P(A \text{ and } B) = (1/4)(1/4) = .0625$
    (d) $P(\text{at least one colorblind}) = 1 - (3/4)(3/4) = .4375$

4.69   Let A = car S starts, and B = car T starts.
   (a) P(both fail to start) = $P(\overline{A}\ and\ \overline{B}\,)$ = (2/10)(3/10) = .06
   (b) P(at least one fails to start) = $1 - P(A\ and\ B)$ = 1 − (8/10)(7/10) = .44
   (c) P(exactly one fails to start) = 1 − [P(both fail to start) + P(both start)]
            = 1 − [(6/100) + (56/100)] = .38

4.71   (a)   Let events A = S arrives at 1:30 and B = T arrives at 1:30.
            $P(A\ and\ B)$ = (1/3)(1/3) = .1111
   (b)   There are 3 times in which they both arrive at the same time.
            P(they arrive at same time) = 3(1/9) = .3333
   (c)   There are two ways in which S arrives 30 minutes before T.
            P(S arrives 30 minutes before T) = 2(1/9) = .2222

4.73   Let events A = husband lives for 25 more years, and B = wife lives for 25 more years.
   $P(A)$ = 7/10, and $P(B)$ = 9/10.
   (a) $P(\overline{A}\,)$ = 1 − (7/10) = .3
   (b) $P(A\ and\ B)$ = (7/10)(9/10) = .63
   (c) P(at least one of the two live for 25 more years)
            = 1 − P(neither one lives for 25 more years) = 1 − (3/10)(1/10) = .97

4.75 $n(S)$ = (400 + 190 + 35 + 75 + 15 + 10 + 5 + 270) = 1000
   (a)   (400 + 75 + 15 + 5)/1000 = .495
   (b)   (190 + 35 + 75 + 15 + 10 + 5)/1000 = .33
   (c)   (400 + 190 + 75 + 15 + 10 + 5)/1000 = .695

4.77   (a)   (40 + 20)/150 = 60/150 = .4
   (b)   (40 + 40 + 5)/150 = .5667
   (c)   Let events A = approve, and B = building and trade.
            $P(A|B)$ = [(40/150)]/[(40 + 10)/150] = .8
   (d)   Let events A = businessmen and B = approve.
            $P(A|B)$ = [(40/150)]/[(40 + 40 + 5)/150] = 40/85 = .4706
   (e)   Let events A = businessmen and B = approve.
            $P(A\ or\ B)$ = [(40 + 20)/150] + [(40 + 40 + 5)/150] − [40/150] = .7
   (f)   Events "businessmen" and "approve" are not independent. To see this, let events
        A = businessmen, and B = approve. $P(A|B)$ = .4706 by part (d) above, but
        $P(A)$ = .4 by part (a) above. So, $P(A|B) \neq P(A)$. The events are not mutually
        exclusive as some businessmen approve of the hotel being built.

4.79   (a)   479,001,600
   (b)   20!/(20 − 3)! = (20)(19)(18) = 6840
   (c)   20!/(3!)(20 − 3)! = (20)(19)(18)/(3)(2)(1) = 1140
   (d)   (100)(99) = 9900

4.81   (a) $_5C_2$ = 5!/(2!)(5 − 2)! = 10. The 10 pairs of elements are
            {A, B}, {A, C}, {A, D}, {A, E}, {B, C}, {B, D}, {B, E}, {C, D}, {C, E}, {D, E}.
   (b) $_5P_2$ = 5!/(5 − 2)! = 20. The 20 ordered pairs of elements are: AB, BA, AC, CA, AD,
            DA, AE, EA, BC, CB, BD, DB, BE, EB, CD, DC, CE, EC, DE, ED.

4.83 $_{12}C_5$ = 12!/(5!)(12 − 5)! = 792

4.85   (a) $(_4C_4)(_{48}C_1)/(_{52}C_5) = (1)(48)/2{,}598{,}960 = .000018$

       (b) $(_{13}C_5)(_{39}C_0)/(_{52}C_5) = (1287)(1)/2{,}598{,}960 = .0005$

       (c) $(_4C_3)(_4C_2)(_{44}C_0)/(_{52}C_5) = (4)(6)(1)/2{,}598{,}960 = .000009$

4.87   The exact probability is $1 - {_{365}P_{35}}/365^{35} = .8144$. In the Resampling Stats simulation of 1000 trials, 2 or more students with the same birthday occurred $1000 - 175 = 825$ times. The simulated probability is $825/1000 = .825$.

Chapter 5

5.1   (a) $E = \{4260, 4261, 4262, 4263, 4264, 4265, 4266, 4267, 4268, 4269\}$

      (b) $E = \{2468, 2648, 4268, 4628, 6248, 6428\}$

5.3 $S = \{0A, 0B, 0C, 0D, 0E, 1A, 1B, 1C, 1D, 1E\}$

      (a) $E = \{0D, 0E\}$ (b) $E = \{1A, 1B, 1C, 1D\}$

      (c) $E = \{0A, 0B, 1A, 1B\}$ (d) $E = \{1A, 1B, 1C, 1D, 1E\}$

5.5   (a) $S = \{0, 1, 2, 3, 4, 5, 6, 7, 8, 9\}$ (b) $E = \{1, 3, 5, 7, 9\}$ $P(E) = 5/10$

      (c) $E = \{7, 8, 9\}$ $P(E) = 3/10$

5.7   Let b = boy and g = girl. Notice that this Exercise is much like Exercise 4.14, with $b = E$ and $g = O$. $S = \{bbbb, bbbg, bbgb, bgbb, gbbb, bbgg, bgbg, bggb, gbbg, gbgb, ggbb, gggb, ggbg, gbgg, bggg, gggg\}$, where the list for a particular family is from youngest to oldest. Now $n(S) = 16$.

      (a)   Let E be the event of two boys.

           $E = \{bbgg, bgbg, bggb, gbbg, gbgb, ggbb\}$ $P(E) = 6/16$

      (b)   Let E be the event the oldest two children are boys

           $E = \{bbbb, bgbb, gbbb, ggbb\}$ $P(E) = 4/16$

      (c)   Let E be the event all are girls. $E = \{gggg\}$ $P(E) = 1/16$

      (d)   Not all are girls is the same as one or more boys. Let E be the event of one or more boys. $P(E) = 15/16$

5.9   (a)   Let E be the event of matching either 0, 1, or 2.

           $P(E) = (1{,}947{,}792 + 2{,}261{,}952 + 883{,}575)/5{,}245{,}786 = .971$

      (b)   Let E be the event of one or more matches.

           $P(E) = (2{,}261{,}952 + 883{,}575 + 142{,}800 + 9450 + 216 + 1)/5{,}245{,}786$

           $= 3{,}297{,}994/5{,}245{,}786 = .629$ Note that an easier way of calculating the numerator is $5{,}245{,}786 - 1{,}947{,}792$.

      (c)   Let E be the event of matching 3 numbers. $P(E) = 142{,}800/5{,}245{,}786 = .0272$

      (d)   Let E be the event of matching 4 or 5 numbers.

           $P(E) = (9450 + 216)/5{,}245{,}786 = .0018$

      (e)   Let E be the event of matching all 6 numbers.

           $P(E) = 1/5{,}245{,}786 = .000000191$, which is a less than 1 in 5 million.

5.11   (a)   Discrete   (b)   Continuous   (c)   Continuous

       (d)   Discrete   (e)   Continuous   (f)   Discrete

5.13   (a)   2, 3, 5, 6, 7, 8, 9   The word and its x value are

          Word: The grass needs cutting and should be cut by tomorrow afternoon

            x: 3   5   5   7   3   6   2 3   2   8   9

      (b)   Discrete

5.15 (a) 0, 1, 2, 3, 4, ... , 99, 100 (b) Discrete

5.17 (a) Continuous $x \geq 0$ (b) Discrete $x = 0, 1, ... , 50$

5.19
x	25	25.7	26.6	27.2	27.3	27.8	28.4	29
$P(x)$	1/15	2/15	1/15	1/15	4/15	2/15	2/15	2/15

5.21
x	1	2	3	4	5	6
$P(x)$	14/25	4/25	2/25	2/25	2/25	1/25

5.23
x	1	2	3	4	5	6	7	8	9
$P(x)$	13/50	13/50	7/50	8/50	4/50	1/50	2/50	1/50	1/50

5.25 $P(1) = 234/825 = .284$ $P(2) = 297/825 = .360$ $P(3) = 294/825 = .356$
x	1	2	3
$P(x)$	.284	.360	.356

5.27

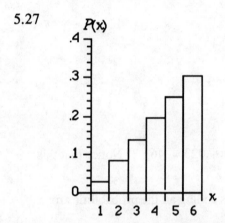

5.29 (a) $\mu = 0(1/10) + 1(2/10) + 2(3/10) + 3(4/10) = 2$
(b) $\sigma^2 = (0-2)^2(1/10) + (1-2)^2(2/10) + (2-2)^2(3/10) + (3-2)^2(4/10) = 1$
(c) $\sigma = \sqrt{1} = 1$
(d) Skewed to the left

5.31 (a) $\mu = 0(1/8) + 1(3/8) + 2(3/8) + 3(1/8) = 3/2 = 1.5$
(b) $\sigma^2 = 0^2(1/8) + 1^2(3/8) + 2^2(3/8) + 3^2(1/8) - (3/2)^2 = .75$
(c) $\sigma = \sqrt{.75} \doteq .866$
(d) Symmetric

5.33 (a) $\mu = 1(2/15) + 2(10/15) + 3(2/15) + 4(1/15) = 32/15 = 2.133$
(b) $\sigma^2 = 1^2(2/15) + 2^2(10/15) + 3^2(2/15) + 4^2(1/15) - (32/15)^2 = .516$
(c) $\sigma = \sqrt{.516} = .718$

5.35 The probability distribution for x is:
x	1	2	3
$P(x)$	1/6	3/6	2/6

$\mu = 1(1/6) + 2(3/6) + 3(2/6) = 13/6 = 2.167$
$\sigma^2 = 1^2(1/6) + 2^2(3/6) + 3^2(2/6) - (13/6)^2 = .472$

5.37 Let x and y be the bus company's gain for the express and local stops respectively. The probability distributions for x and y respectively are

x	-500	5500
$P(x)$	.30	.70

y	-750	9250
$Q(y)$	.50	.50

$E(x) = (-500)(.30) + (5500)(.70) = \$3700$ $E(y) = (-750)(.50) + (9250)(.50) = \$4250$
The company should bid on the local stops contract, as the expected gain is larger.

5.39 Let x be the amount you win. The probability distribution for x is

x	-1	0	74	1499	$999,999$
$P(x)$	.97093534	.02722185	.00180145	.00004118	.00000019

$E(x) = -1(.97093534) + 0(.02722185) + 74(.00180145) + 1499(.00004118)$
$+ 999999(.00000019) = -.586$. Your expected loss is 58.6 cents per ticket bought.

5.41 (a) Yes (b) No; the number of trials is not determined in advance
(c) No; the trials are dependent (d) Yes (e) No; the trials are dependent

5.43 (a) $P(0) = \dfrac{4!}{0!(4-0)!}(.6)^0(.4)^{4-0} = (.4)^4 = .026$

(b) $P(1) = \dfrac{4!}{1!(4-1)!}(.6)^1(.4)^{4-1} = 4(.6)^1(.4)^3 = .154$

(c) $P(2) = \dfrac{4!}{2!(4-2)!}(.6)^2(.4)^{4-2} = 6(.6)^2(.4)^2 = .346$

5.45 (a) $P(0) + P(1) = .004 + .027 = .031$
(b) $P(6) + P(7) = .147 + .070 = .217$
(c) $P(5) + P(6) + P(7) + P(8) = .221 + .147 + .070 + .023 = .461$ Notice that this differs from part (b).
(d) $P(6) = .147$
(e) $1 - P(0) = 1 - .004 = .996$ This is the same as adding all the probabilities for x greater than zero.

5.47 Assume a binomial experiment with $n = 20$ and $p = .6$. The number of successes x is the number opposed to the proposed development.
(a) $P(10) = .117$
(b) $P(\text{more than } 13) = .124 + .075 + .035 + .012 + .003 + (0+) + (0+) = .249$
(c) $P(\text{less than } 10)$
$= (0+) + (0+) + (0+) + (0+) + (0+) + .001 + .005 + .015 + .035 + .071 = .127$

5.49 (a) .092 (b) $.057 + .103 + .147 + .171 = .478$

5.51 Assume a binomial experiment with $n = 19$ and $p = .10$. The number of successes x is the number of Caucasians aged 45-74 that have diabetes.
(a) $P(3) = .180$
(b) $P(0) + P(1) = .135 + .285 = .420$

5.53 Assume a binomial experiment with $n = 15$ and $p = .20$. The number of successes x is the number of Hispanics aged 45-74 that have diabetes.
(a) $P(3) = .250$
(b) $P(0) + P(1) = .035 + .132 = .167$

5.55   Assume a binomial experiment with $n = 18$ and $p = .25$. The number of successes x is the number of résumés that contain a major fabrication.

$$P(5) = \frac{18!}{5!(13)!}(.25)^5(.75)^{13} = .1988$$

5.57   Assume a binomial experiment with $n = 17$ and $p = .36$. The number of successes x is the number of students that report being frequently bored in class during their last year in high school.

$$P(3) = \frac{17!}{3!(14)!}(.36)^3(.64)^{14} = .0614$$

5.59   Assume a binomial experiment with $n = 405$ and $p = .154$. The number of successes x is the number of black teachers.
   (a) $\mu = (405)(.154) = 62.37$
   (b) $\sigma = \sqrt{(405)(.154)(.846)} = 7.26$
   (c) $z = \dfrac{15 - (405)(.154)}{\sqrt{(405)(.154)(.846)}} = -6.52$, or 6.52 standard deviations
   (d)   Now use $p = .057$ $\mu = (405)(.057) = 23.09$ $\sigma = \sqrt{(405)(.057)(.943)} = 4.67$
          $z = \dfrac{15 - (405)(.057)}{\sqrt{(405)(.057)(.943)}} = -1.73$, or 1.73 standard deviations   (e)   Probably not

5.61   (a) $\mu = (50)(.07) = 3.5$
   (b) $\sigma^2 = (50)(.07)(.93) = 3.255$
   (c) $\sigma = \sqrt{(50)(.07)(.93)} = 1.804$

5.63

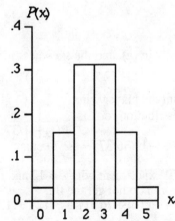

Each distribution is symmetric. In general, for any n, the binomial distribution is symmetric when $p = .5$.

5.65   (a)   The blood cholesterol level   Continuous
   (b)   The population (conceptual) is the collection of all blood cholesterol level scores of people who will use the diet. The sample is the collection of 50 readings.

5.67   (a) $\mu = 2(4/6) + 3(1/6) + 13(1/6) = 4$

(b) $\sigma^2 = 2^2(4/6) + 3^2(1/6) + (13)^2(1/6) - 4^2 = 98/6 = 16.333$
(c) $\sigma = \sqrt{98/6} = 4.041$

5.69 (a)   We need $\Sigma P(x) = 1$, and $2/12 + 5/12 + 2/12 = 9/12$. $P(3) = 1 - 9/12 = 3/12 = .25$
(b) $\mu = 1(2/12) + 2(5/12) + 3(3/12) + 6(2/12) = 33/12 = 2.75$
(c) $\sigma^2 = 1^2(2/12) + 2^2(5/12) + 3^2(3/12) + 6^2(2/12) - (33/12)^2 = 2.521$
(d) $\sigma = \sqrt{363/144} = 1.588$

5.71 $\Sigma P(x) = c + 2c = 3c = 1$ Therefore, $c = 1/3$.

5.73 Note there are 6 possible outcomes: 123, 132, 213, 231, 312, and 321 with x assuming
values 3, 1, 1, 0, 0, and 1 respectively.
(a)

x	0	1	3
$P(x)$	2/6	3/6	1/6

(b) $\mu = 0(2/6) + 1(3/6) + 3(1/6) = 1$ So the average value should be about 1.

5.75 $q^4 = P(0) = .2401$, so $q = (.2401)^{1/4} = .7$
$p = 1 - .7 = .3$ $n = 4$
(a) $\mu = 4(.3) = 1.2$
(b) $\sigma^2 = 4(.3)(.7) = .84$
(c) $\sigma = \sqrt{.84} = .917$

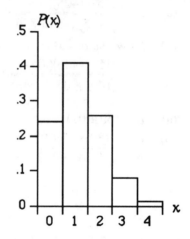

5.77 Notice the symmetry in (a), and the skewness in (b) and (c).
(a) $p = .5$ (b) $p = .8$ (c) $p = .3$

5.79 Let x be the amount the player wins.
The probability distribution for x is

x	35	−1
$P(x)$	1/37	36/37

$E(x) = (35)(1/37) + (-1)(36/37) = -2.7$ cents

5.81 Assume a binomial experiment with $n = 12$ and $p = .8$. The number of successes x is the
number of people who do not get the flu.
(a) $P(8) + P(9) + P(10) + P(11) + P(12) = .133 + .236 + .283 + .206 + .069 = .927$
(b) $P(10) = .283$

5.83 Assume a binomial experiment with $n = 7$ and $p = .7$. The number of successes x is the
number of complaints investigated in the sample.
(a) $P(7) = .082$ (b) $P(1) = .004$ (c) P(no more than 2) = (0+) + .004 + .025 = .029

5.85 Assume a binomial experiment with $n = 12$ and $p = .99$. The number of successes x is the
number of the 12 components that do not break down within the first year.

P(none of the components break down in the first year) = $P(12) = (.99)^{12} = .886$

5.87   (a) $\mu = 12(.8) = 9.6$ (b) $\sigma^2 = 12(.8)(.2) = 1.92$ (c) $\sigma = \sqrt{1.92} = 1.386$

5.89   (a) $\mu = 7(.7) = 4.9$ (b) $\sigma^2 = 7(.7)(.3) = 1.47$ (c) $\sigma = \sqrt{1.47} = 1.212$

5.91 $343/2500 = .1372$ It is not surprising that a name occurred three times, as it will happen 13-14 percent of the time.

Chapter 6

6.1   Notice that the height of the rectangle is 1/2.
    (a) $(1/2)(1/2) = .25$
    (b) $[(7/4) - (1/4)](1/2) = .75$
    (c) $[2 - (3/5)](1/2) = .7$, or another way: $1 - P(x < 3/5) = 1 - (3/5)(1/2) = .7$
    (d) $P(x < 1/5 \text{ or } x > 2/5) = 1/10 + (1 - 2/10) = .9$
    (e) $2(1/2) = 1$ Notice that the event $(x < 2)$ is certain to occur.
    (f) $P(x = 3/4) = 0$ Notice that x is a continuous random variable, and so the probability that x will assume a particular value is 0.

6.3   (a) $P(x > 700) + P(x < 300) = 2P(x > 700) = 2[.5 - (.3413 + .1359)] = .0456$
    (b) $P(x > 500) = 1/2$ Notice the distribution is symmetric, and so the total probability of 1 is split evenly below and above the mean.
    (c) $P(600 < x < 700) = .1359$
    (d) $P(x < 600) = .5 + .3413 = .8413$
    (e) $P(300 < x < 500) = .1359 + .3413 = .4772$
    (f) $P(400 < x < 600) = .3413 + .3413 = .6826$
    (g) $P(300 < x < 700) = .1359 + .3413 + .3413 + .1359 = .9544$

6.5   (a) $P(z < 1.41) = .5 + \text{Area from 0 to } 1.41 = .5 + .4207 = .9207$
    (b) $P(z < -1.72) = .5 - \text{Area from 0 to } 1.72 = .5 - .4573 = .0427$
    (c) $P(z > 1.51) = .5 - \text{Area from 0 to } 1.51 = .5 - .4345 = .0655$
    (d) $P(z > -2.43) = .5 + \text{Area from 0 to } 2.43 = .5 + .4925 = .9925$

6.7   (a) $P(z < 0) = P(z > 0) = .5$
    (b) $P(-.67 < z < 0) = P(0 < z < .67) = .2486$
    (c) $P(-2.3 < z < -1.45) = P(0 < z < 2.3) - P(0 < z < 1.45) = .4893 - .4265 = .0628$
    (d) $P(-.73 < z < 2.31) = P(0 < z < .73) + P(0 < z < 2.31) = .4896 + .2673 = .7569$
    (e) $P(z < 1.96) = .5 + P(0 < z < 1.96) = .5 + .475 = .975$
    (f) $P(-1 < z < 1) = 2P(0 < z < 1) = 2(.3413) = .6826$
    (g) $P(-3 < z < 3) = 2P(0 < z < 3) = 2(.4987) = .9974$

6.9   (a) $P(z < c) = .9772$, or $P(0 < z < c) = .9772 - .5 = .4772$ $c = 2$
    (b) $P(z < c) = .0668$, or $P(0 < z < -c) = .5 - .0668 = .4332$ $c = -1.5$
    (c) $P(z > c) = .5$ $c = 0$
    Notice that the standard normal distribution is symmetric about $z = 0$.
    (d) $P(z > c) = .9599$, or $P(0 < z < -c) = .5 - (1 - .9599) = .4599$ $c = -1.75$

6.11   (a) $P(z < c) = .9573$, or $P(0 < z < c) = .9573 - .5 = .4573$ $c = 1.72$
    (b) $P(z < c) = .1075$, or $P(0 < z < -c) = .5 - .1075 = .3925$ $c = -1.24$
    (c) $P(z > c) = .0793$, or $P(0 < z < c) = .5 - .0793 = .4207$ $c = 1.41$
    (d) $P(z \geq c) = .9929$, or $P(0 < z < -c) = .5 - (1 - .9929) = .4929$ $c = -2.45$

(e) $z_{.02} = c$, where $P(0 < z < c) = .5 - .02 = .48$ $c = 2.05$

(f) $z_{.75} = c$, where $P(z > c) = .75$, or $P(0 < z < -c) = .25$ $c = -.67$

(g) $z_{.90} = c$, where $P(z > c) = .90$, or $P(0 < z < -c) = .40$ $c = -1.28$

(h) $z_{.35} = c$, where $P(0 < z < c) = .5 - .35 = .15$ $c = .39$

6.13

(a) $P(x > 46) = P(z > 0) = .5$, or 50%

(b) $P(x > 50) = P(z > 1) = .5 - .3413 = .1587$, or 15.87%

(c) $P(x > 40) = P(z > -1.5) = .5 + .4332 = .9332$, or 93.32%

(d) $P(x < 38) = P(z < -2) = .5 - .4772 = .0228$, or 2.28%

(e) $P(x < 49) = P(z < .75) = .5 + .2734 = .7734$, or 77.34%

(f) $P(45 < x < 49) = P(-.25 < z < .75) = .0987 + .2734 = .3721$, or 37.21%

(g) $P(50 < x < 54) = P(1 < z < 2) = .4772 - .3413 = .1359$, or 13.59%

(h) $P(x < 46 \text{ or } x > 56) = P(z < 0) + P(z > 2.5) = .5 + .0062) = .5062$ or 50.62%

(i) $P(-1.5 < z < 1.5) = 2(.4332) = .8664$, or 86.64%

(j) $P(z < -2.3 \text{ or } z > 2.3) = 1 - 2(.4893) = .0214$, or 2.14%

6.15

(a) $Q_1 = 150 - (.67)(10) = 143.30$

(b) $P_{65} = 150 + (.39)(10) = 153.90$

(c) $z = (165 - 150)/10 = 1.5$, and $P(z < 1.5) = .9332$ Percentile rank = 93.32

(d) $z = (145 - 150)/10 = -.5$, and $P(z < -.5) = .3085$ Percentile rank = 30.85

6.17 $A = P(z < -1.83) = .0336$ $B = P(-1.83 < z < .6) = .6922$

         $C = P(.6 < z < 2.1) = .2563$ $D = P(z > 2.1) = .0179$

6.19

(a) $500 + (1.65)(100) = 665$

(b) $500 \pm (.52)(100) = 500 \pm 52$, or (448, 552)

(c)   Let x be a test score. $P(x > 650) = P(z > 1.5) = .5 - .4332 = .0668$. Of the 1000 new students, about $(1000)(.0668) = 67$ are expected to score more than 650.

6.21

(a)   Notice that 3 years is 36 months. Now $P(x < 36) = P(z < -1.33) = .0918$, so 9.18% of the sensors will fail the guarantee.

(b) $c = 48 + (-2.33)(9) = 27$ months

6.23   Let x be diastolic blood pressure measured in mm.

(a) $P(x < 90) = P\left(z < \dfrac{90 - 79}{11} = 1\right) = .5 + .3413 = .8413$

(b) $P(x \geq 90) = 1 - P(x < 90) = 1 - .8413 = .1587$

6.25

(a) $(0+) + .003 + .015 + .047 + .101 + .162 + .198 + .189 = .715$

(b) $\mu = (16)(.4) = 6.4$, and $\sigma = \sqrt{(16)(.4)(.6)}$

$$P(x \leq 7) \doteq P\left(z \leq \frac{7.5 - 6.4}{\sqrt{(16)(.4)(.6)}} = .56\right) = .5 + .2123 = .7123$$

6.27

(a)   .016

(b) $\mu = (12)(.5) = 6$, and $\sigma = \sqrt{(12)(.5)(.5)} = 1.73$

$$P(x = 10) \doteq P\left(\frac{9.5 - 6}{\sqrt{(12)(.5)(.5)}} = 2.02 < z < \frac{10.5 - 6}{\sqrt{(12)(.5)(.5)}} = 2.60\right)$$

$$= .4953 - .4783 = .0170$$

6.29 (a) $.024 + .061 + .118 = .203$

(b) $\mu = (15)(.6) = 9$, and $\sigma = \sqrt{(15)(.6)(.4)}$

$$P(5 \le x \le 7) \doteq P\left(\frac{4.5 - 9}{\sqrt{(15)(.6)(.4)}} = -2.37 \le z \le \frac{7.5 - 9}{\sqrt{(15)(.6)(.4)}} = -.79\right)$$

$$= .4911 - .2852 = .2059$$

6.31 Now $n = 300$ and $p = .76$ So $\mu = (300)(.76) = 228$, and $\sigma = \sqrt{(300)(.76)(.24)}$

$$P(x \ge 211) \doteq P\left(z \ge \frac{210.5 - 228}{\sqrt{(300)(.76)(.24)}} = -2.37\right) = .5 + .4911 = .9911$$

6.33 Now $n = 300$ and $p = .41$ So $\mu = (300)(.41) = 123$, and $\sigma = \sqrt{(300)(.41)(.59)}$

$$P(x \le 99) \doteq P\left(z \le \frac{99.5 - 123}{\sqrt{(300)(.41)(.59)}} = -2.76\right) = .5 - .4971 = .0029$$

6.35 Now $n = 500$ and $p = .65$ So $\mu = (500)(.65) = 325$, and $\sigma = \sqrt{(500)(.65)(.35)}$

(a) $P(305 \le x \le 345) \doteq P\left(\frac{304.5 - 325}{\sqrt{(500)(.65)(.35)}} = -1.92 \le z \le \frac{345.5 - 325}{\sqrt{(500)(.65)(.35)}} = 1.92\right)$

$$= 2(.4726) = .9452$$

(b) $P(x \le 295) \doteq P\left(z \le \frac{295.5 - 325}{\sqrt{(500)(.65)(.35)}} = -2.77\right) = .5 - .4972 = .0028$

If the treatment had no effect, only about 3 in 1000 times, on average, would the number of deaths be less than or equal to 295. The evidence suggests that the figure 65% may be high.

6.37 (a) Now $n = 405$ and $p = .154$ So $\mu = (405)(.154)$ and $\sigma = \sqrt{(405)(.154)(.846)}$

$$P(x \le 15) \doteq P\left(z \le \frac{15.5 - (405)(.154)}{\sqrt{(405)(.154)(.846)}} = -6.45\right) \doteq 0$$

(b) Now $\mu = (405)(.057)$ and $\sigma = \sqrt{(405)(.057)(.943)}$

$$P(x \le 15) \doteq P\left(z \le \frac{15.5 - (405)(.057)}{\sqrt{(405)(.057)(.943)}} = -1.63\right) = .5 - .4484 = .0516$$

6.39 (a) $\mu = 1(1/6) + 2(2/6) + 3(3/6) = 7/3$ $\sigma^2 = 1^2(1/6) + 2^2(2/6) + 3^2(3/6) - (7/3)^2 = 5/9$

(b) The largest value the sample mean can be is 3, and this can occur only by getting 49 3's. Similarly, the smallest value is 1, which can occur only if all 49 values are 1.

(c) $\mu_{\bar{x}} = \mu_x = 7/3$, and $\sigma_{\bar{x}}^2 = \dfrac{(5/9)}{49} = \dfrac{5}{441}$

(d) Approximately normal according to the Central Limit Theorem

6.41 (a) $\mu_{\bar{x}} = \mu_x = 200$

(b) $\sigma_{\bar{x}} = \dfrac{100}{\sqrt{100}} = 10$

In parts (c) and (d), use $z = \dfrac{\bar{x} - 200}{10}$

(c) $P(195 < \bar{x} < 205) = P(-.5 < z < .5) = 2(.1915) = .3830$

(d) $P(\bar{x} > 210) = P(z > 1) = .5 - .3413 = .1587$

In parts (e) and (f), use $z = \dfrac{x - 200}{100}$

(e) $P(195 < x < 205) = P(-.05 < z < .05) = 2(.0199) = .0398$

(f) $P(x > 210) = P(z > .1) = .5 - .0398 = .4602$

6.43 (a) $\mu_{\bar{x}} = \mu_x = 200$

(b) $\sigma_{\bar{x}} = \dfrac{100}{\sqrt{400}} = 5$

(c) $P(195 < \bar{x} < 205) = P(-1 < z < 1) = 2(.3413) = .6826$

(d) $P(\bar{x} > 210) = P(z > 2) = .5 - .4772 = .0228$

6.45 Now $\mu_{\bar{x}} = \mu_x = 510$, and $\sigma_{\bar{x}} = \dfrac{90}{\sqrt{100}} = 9$

(a) $P(\bar{x} > 530) = P(z > 2.22) = .5 - .4868 = .0132$

(b) $P(\bar{x} < 500) = P(z < -1.11) = .5 - .3665 = .1335$

(c) $P(495 < \bar{x} < 515) = P(-1.67 < z < .56) = .4525 + .2123 = .6648$

6.47 (a) $\mu_{\bar{x}} = \mu_x = 6$ and $\sigma_{\bar{x}} = \dfrac{1}{\sqrt{100}} = .1$

Then $P(\bar{x} > 5.9) = P(z > -1) = .5 + .3413 = .8413$

(b) $P(\bar{x} < 5.75) = P(z < -2.5) = .5 - .4938 = .0062$

(c) We have to find W with $P(\bar{x} < W) = .01$, with the mean weight being 6 ounces and a standard deviation of 1 ounce. $W = \mu_{\bar{x}} + z\sigma_{\bar{x}} = 6 + (-2.33)(.1) = 5.767$

6.49 (a) $P(x < 1) = 1 - (.2031 + .1455 + .1043 + .0747 + .0535 + .1354) = .2835$

(b) $P(x > 6) = .1354$

(c) $P(1 < x < 3) = .2031 + .1455 = .3486$

(d) $P(x > 1) = 1 - .2835 = .7165$

(e) $P(x < 2 \text{ or } x > 3) = 1 - .1455 = .8545$

6.51 (a) $P(0 < z < 1.86)$ = Area from 0 to 1.86 = .4686

(b) $P(-1.48 < z < 0)$ = Area from 0 to 1.48 = .4306

(c) $P(-1.02 < z < 2.74)$ = Area from 0 to 1.02 + Area from 0 to 2.74
= .3461 + .4969 = .8430

(d) $P(1.09 < z < 1.34)$ = Area from 0 to 1.34 − Area from 0 to 1.09
= .4099 − .3621 = .0478

(e) $P(-.99 < z < -.30)$ = Area from 0 to .99 − Area from 0 to .30
= .3389 − .1179 = .2210

6.53 (a) $P(z < 1.71)$ = .5 + Area from 0 to 1.71 = .5 + .4564 = .9564

(b) $P(z < -1.92)$ = .5 − Area from 0 to 1.92 = .5 − .4726 = .0274

(c) $P(z > 1.11)$ = .5 − Area from 0 to 1.11 = .5 − .3665 = .1335

(d) $P(z > -2.55)$ = .5 + Area from 0 to 2.55 = .5 + .4946 = .9946

6.55 (a) $P(38 < x < 47) = P(-1 < z < .5) = .3413 + .1915 = .5328$

(b) $P(48 < x < 54) = P(.67 < z < 1.67) = .4525 - .2486 = .2039$

(c) $P(x > 59) = P(z > 2.5) = .5 - .4938 = .0062$
(d) $P(x < 52) = P(z < 1.33) = .5 + .4082 = .9082$

6.57 (a) $P_{10} = 180 + (-1.28)(20) = 154.4$
(b) $P_{95} = 180 + (1.65)(20) = 213$
(c) $P(x < 190) = P(z < .5) = .5 + .1915 = .6915$, so the percentile rank is 69.15
(d) $P(x < 140) = P(z < -2) = .0228$, so the percentile rank is 2.28

6.59 (a) $P(x < 180) = P(z < .8) = .5 + .2881 = .7881$, or 78.81%
(b) We need to find T where $P(x > T) = .05$.
So $T = \mu + z\sigma = 160 + (1.65)(25) = 201.25$ minutes.

6.61 (a) .196
(b) $\mu = 15(.5) = 7.5$ and $\sigma = \sqrt{15(.5)(.5)}$

$$P(x = 7) \doteq P\left(\frac{6.5 - 7.5}{\sqrt{(15)(.5)(.5)}} = -.52 < z < \frac{7.5 - 7.5}{\sqrt{(15)(.5)(.5)}} = 0 \right) = .1985$$

6.63 (a) $.004 + .012 + .031 + .065 + .114 = .226$
(b) $\mu = 20(.7) = 14$ and $\sigma = \sqrt{20(.7)(.3)}$

$$P(8 \le x \le 12) \doteq P\left(\frac{7.5 - 14}{\sqrt{(20)(.7)(.3)}} = -3.17 \le z \le \frac{12.5 - 14}{\sqrt{(20)(.7)(.3)}} = -.73 \right)$$
$$= .4992 - .2673 = .2319$$

6.65 (a) $.142 + .189 + .198 + .162 + .101 + .047 + .015 + .003 + (0+) = .857$
(b) $\mu = (16)(.6) = 9.6$ and $\sigma = \sqrt{16(.6)(.4)}$

$$P(x \ge 8) \doteq P\left(z \ge \frac{7.5 - 9.6}{\sqrt{(16)(.6)(.4)}} = -1.07 \right) = .5 + .3577 = .8577$$

6.67 In each part, use the following: $\mu = 400(.5) = 200$, and $\sigma = \sqrt{400(.5)(.5)} = 10$.

(a) $P(x \ge 210) \doteq P\left(z \ge \frac{209.5 - 200}{10} = .95 \right) = .5 - .3289 = .1711$

(b) $P(180 \le x \le 220) \doteq P\left(\frac{179.5 - 200}{10} = -2.05 \le z \le \frac{220.5 - 200}{10} = 2.05 \right)$
$= 2(.4798) = .9596$

(c) $P(170 \le x \le 230) \doteq P\left(\frac{169.5 - 200}{10} = -3.05 \le z \le \frac{230.5 - 200}{10} = 3.05 \right)$
$= 2(.4989) = .9978$

(d) $P(x \le 184) \doteq P\left(z \le \frac{184.5 - 200}{10} = -1.55 \right) = .5 - .4394 = .0606$

(e) $P(x = 200) \doteq P\left(\frac{199.5 - 200}{10} = -.05 < z < \frac{200.5 - 200}{10} = .05 \right) = 2(.0199) = .0398$

(f) $P(x = 210) \doteq P\left(\frac{209.5 - 200}{10} = .95 < z < \frac{210.5 - 200}{10} = 1.05 \right)$
$= .3531 - .3289 = .0242$

6.69 $\mu = 400(.154) = 61.6$, and $\sigma = \sqrt{400(.154)(.846)}$

$$P(x \le 69) \doteq P\left(z \le \frac{69.5 - 61.6}{\sqrt{(400)(.154)(.846)}} = 1.09\right) = .5 + .3621 = .8621$$

6.71   (a) $\mu_{\bar{x}} = \mu_x = 71$ (b) $\sigma_{\bar{x}} = \dfrac{2.5}{\sqrt{100}} = .25$

         (c) $P(\bar{x} > 70.5) = P(z > -2) = .5 + .4772 = .9772$, or 97.72%

         (d) $P(x > 70.5) = P(z > -.2) = .5 + .0793 = .5793$, or 57.93%

6.73   (a) $P(x < 126) = P(z < -2) = .5 - .4772 = .0228$, or 2.28%

         (b) $P(x > 129) = P(z > 1) = .5 - .3413 = .1587$, or 15.87%

         (c) $P(127.5 < x < 130.5) = P(-.5 < z < 2.5) = .1915 + .4938 = .6853$, or 68.53%

         (d)   Now $\mu_{\bar{x}} = \mu_x = 128$, and $\sigma_{\bar{x}} = \dfrac{1}{\sqrt{25}} = .2$

             $P(\bar{x} < 127.4) = P(z < -3) = P(z > 3) = .5 - P(0 < z < 3) = .5 - .4987 = .0013$

         (e)   There is strong evidence to suggest that the true mean is smaller than 128. If the population mean were 128, the chances of observing a sample mean of 127.4 or smaller are only about 13 in 10,000.

Chapter 7

7.1   The maximum error of the estimate is given by $E = z_{\alpha/2}\left(\dfrac{s}{\sqrt{n}}\right)$, and $\bar{x} \pm E$ is the form of the confidence interval.

         (ai) $(1.96)(16/\sqrt{64}) = 3.92$ (aii) 125 ± 3.92 or $121.08 < \mu < 128.92$

         (bi) $(2.58)(25/\sqrt{100}) = 6.45$ (bii) 206 ± 6.45 or $199.55 < \mu < 212.45$

         (ci) $(1.65)(3/\sqrt{81}) = .55$ (cii) $154 \pm .55$ or $153.45 < \mu < 154.55$

         (di) $(1.96)(50/\sqrt{225}) = 6.53$ (dii) 309 ± 6.53 or $302.47 < \mu < 315.53$

         (ei) $(2.58)(7/\sqrt{49}) = 2.58$ (eii) 40 ± 2.58 or $37.42 < \mu < 42.58$

         (fi) $(1.65)(6/\sqrt{144}) = .83$ (fii) $78 \pm .83$ or $77.17 < \mu < 78.83$

7.3 $3.1 \pm (1.96)(1.6/\sqrt{45})$ or $2.63 < \mu < 3.57$

7.5 $.72 \pm (1.65)(.31/\sqrt{33})$ or $.63 < \mu < .81$

7.7   (a) $29,400 \pm (1.96)(6325/\sqrt{40})$ or $27,439.86 < \mu < 31,360.14$

         (b)   Yes, as the confidence interval does not include 25,300.

7.9   (a) $65 \pm (1.65)(6/\sqrt{40})$ or $63.43 < \mu < 66.57$

         (b) $E = (1.65)(6/\sqrt{40}) = 1.57$

         (c)   Yes, since the value of the upper limit of the confidence interval (66.57) is below 80.

7.11   (a) $150 \pm (1.65)(15/\sqrt{50})$ or $146.50 < \mu < 153.50$

         (b) $150 \pm (2.58)(15/\sqrt{50})$ or $144.53 < \mu < 155.47$

         (c)   Shorter than the 99% confidence interval, and longer than the 90%

7.13   (a) $23.9\% \pm (1.96)(4.6\% /\sqrt{34})$ or $22.35\% < \mu < 25.45\%$

      (b) $(1.96)(4.6\% /\sqrt{34}) = 1.55\%$

7.15 $n = \left[\dfrac{(1.65)(1.75)}{.5}\right]^2 = 33.35$ The sample size should be at least 34.

7.17   (a) $n = \left[\dfrac{(1.65)(20)}{4}\right]^2 = 68.06$ Choose a sample of at least size 69.

      (b)   Larger, as more confidence is required. Mathematically, "1.65" in part (a) would be replaced by "1.96"

      (c)   Larger, as more precision is required. Mathematically, "4" in part (a) would be replaced by "2"

7.19   (a) $\bar{x} = 3954.6/37 = 106.881$ $s = \sqrt{\dfrac{37(442,887.6) - (3954.6)^2}{37(36)}} = 23.697$

         $106.881 \pm (1.96)(23.697 /\sqrt{37})$ or $99.245 < \mu < 114.517$

      (b)   Yes, as the confidence interval is above 88

7.21   As 120 is above the confidence interval, the administrator can conclude that mean birthweight at his hospital is smaller than that nationwide, on average.

7.23 $\bar{x} = 1718/39 = 44.05$ $s = \sqrt{\dfrac{39(78,398) - (1718)^2}{39(38)}} = 8.46$

      $44.05 \pm (1.65)(8.46/\sqrt{39})$ or $41.81 < \mu < 46.29$
      The owner will not continue to carry the model, as the mean repair time appears to be more than 41.81 minutes.

7.25   (ai) $H_0: \mu = 72$; $H_a: \mu > 72$ (aii)   Right tailed

      (aiii) Type I: The data suggest that the mean amount of rainfall per year is more than 72, when it is not.
             Type II: The data do not suggest that the mean amount of rainfall per year is more than 72, when it is.

      (bi) $H_0: \mu = 250$; $H_a: \mu \neq 250$ (bii)   Two tailed
      (biii) Type I: The data suggest that the mean number of books borrowed per day is not 250, when it is.
             Type II: The data do not suggest that the mean number of books borrowed per day is different from 250, when in fact it is different.

      (ci) $H_0: \mu = 78$; $H_a: \mu < 78$ (cii)   Left tailed
      (ciii) Type I: The data suggest that the mean July temperature is less than 78, when it is not.
             Type II: The data do not suggest that the mean July temperature is less than 78, when it is.

      (di) $H_0: \mu = 36$; $H_a: \mu > 36$ (dii)   Right tailed
      (diii) Type I: The data suggest that the mean age is more than 36, when it is not.
             Type II: The data do not suggest that the mean age is more than 36, when it is.

      (ei) $H_0: \mu = 29,500$; $H_a: \mu < 29,500$ (eii)   Left tailed
      (eiii) Type I: The data suggest that the mean salary is less than \$29,500, when it is not.
             Type II: The data do not suggest that the mean salary is less than \$29,500, when it is.

      (fi) $H_0: \mu = 18,000$; $H_a: \mu \neq 18,000$ (fii)   Two tailed

(fiii) Type I: The data suggest that the mean number of families below poverty level is not 18,000, when it is.
Type II: The data do not suggest that the mean number of families below poverty level is different from 18,000, when in fact it is different.

(gi) $H_0: \mu = 2.3$; $H_a: \mu > 2.3$ (gii)   Right tailed

(giii) Type I: The data suggest that the mean grade point average is more than 2.3, when it is not.
Type II: The data do not suggest that the mean grade point average is more 2.3, when it is.

7.27   (a) $H_0: \mu = 90$; $H_a: \mu > 90$ $z \geq 1.65$

       (b) $H_0: \mu = 90$; $H_a: \mu \neq 90$ $z \leq -2.58$ or $z \geq 2.58$

       (c) $H_0: \mu = 90$; $H_a: \mu < 90$ $z \leq -1.28$

7.29 $H_0: \mu = 15$; $H_a: \mu > 15$ $\alpha = .05$ Observed value: $z = \dfrac{17.3 - 15}{5.4 / \sqrt{50}} = 3.01$

Critical value: $z = 1.65$ Decision: Reject H_0
The data suggest that the mean depression level is more than 15.

7.31 $H_0: \mu = 500$; $H_a: \mu < 500$ $\alpha = .01$ Observed value: $z = \dfrac{486 - 500}{53 / \sqrt{32}} = -1.49$

Critical value: $z = -2.33$ Decision: Do not reject H_0
The data do not suggest that the mean depth is less than 500 feet.

7.33   (a) $H_0: \mu = 700$; $H_a: \mu \neq 700$ $\alpha = .05$ Observed value: $z = \dfrac{675 - 700}{77 / \sqrt{48}} = -2.25$

       Critical values: $z = \pm 1.96$ Decision: Reject H_0
       The data suggest that the mean number of hours is not 700.

       (b)   Type I   (c) $\alpha = .05$

7.35 $H_0: \mu = 150$; $H_a: \mu > 150$ $\alpha = .05$ Observed value: $z = \dfrac{153 - 150}{7.5 / \sqrt{40}} = 2.53$

Critical value: $z = 1.65$ Decision: Reject H_0 Type I
If an error has been made, then the data suggest that the mean time is more than 150 minutes, when it is not.

7.37 $H_0: \mu = 6.5$; $H_a: \mu < 6.5$ $\alpha = .05$ Observed value: $z = \dfrac{5.84 - 6.5}{2.41 / \sqrt{38}} = -1.69$

Critical value: $z = -1.65$ Decision: Reject H_0 At the 5% level, the evidence supports the hypothesis that the mean recovery time is less than 6.5 days, on average.

7.39 P-value $= 2P(z \geq 3.5) = 2(.5 - .4998) = .0004$ $(< .05)$ Reject H_0

7.41 P-value $= P(z \geq 1.2) = .5 - .3849 = .1151$ $(> .05)$ Do not reject H_0

7.43 P-value $= P(z \leq -2.95) = .5 - .4984 = .0016$ $(< .05)$ Reject H_0

7.45   In parts (a) – (c), the hypotheses are: $H_0: \mu = 30$; $H_a: \mu > 30$

       (a) P-value $= P(\bar{x} \geq 35) = P\left(z \geq \dfrac{35 - 30}{40 / \sqrt{64}} = 1 \right) = .5 - .3413 = .1587$

(b) $P\text{-value} = P(\bar{x} \geq 37) = P\left(z \geq \dfrac{37-30}{40/\sqrt{64}} = 1.40\right) = .5 - .4192 = .0808$

(c) $P\text{-value} = P(\bar{x} \geq 39) = P\left(z \geq \dfrac{39-30}{40/\sqrt{64}} = 1.80\right) = .5 - .4641 = .0359$

(d)   The hypotheses are: $H_0: \mu = 30;\ H_a: \mu \neq 30$
Double the answers in parts (a), (b), (c)
Therefore, the corresponding P-v alues are:
$2(.1587) = .3174 \qquad 2(.0808) = .1616 \qquad 2(.0359) = .0718$

7.47 $H_0: \mu = 500;\ H_a: \mu < 500 \qquad \alpha = .01 \qquad$ Observed value: $z = \dfrac{486-500}{53/\sqrt{32}} = -1.49$
$P\text{-value} = P(z \leq -1.49) = .5 - .4319 = .0681$, which is larger than .01
Decision: Do not reject H_0
The data do not suggest that the mean depth is less than 500 feet.

7.49 $H_0: \mu = 15;\ H_a: \mu > 15 \qquad \alpha = .05 \qquad$ Observed value: $z = \dfrac{17.3-15}{5.4/\sqrt{50}} = 3.01$
$P\text{-value} = P(z \geq 3.01) = .5 - .4987 = .0013$, which is less than .05
Decision: Reject H_0 The data suggest that the mean depression level is more than 15.

7.51 $H_0: \mu = 6;\ H_a: \mu \neq 6 \qquad$ (a)   Observed value: $z = \dfrac{5.95-6}{.15/\sqrt{40}} = -2.11$
$\qquad P\text{-value} = 2P(z \leq -2.11) = 2[.5 - .4826] = .0348$
(bi)   Yes, as .0348 < .10   (bii)   Yes, as .0348 < .05   (biii)   No, as .0348 > .01
(ci)   Type I   (cii)   Type I   (ciii)   Type II

7.53 $H_0: \mu = 64;\ H_a: \mu \neq 64 \qquad$ Observed value: $z = -2.38$
(a) $P\text{-value} = 2P(z \leq -2.38) = 2(.5 - .4913) = .0174$
(b)   All levels of significance greater than or equal to .0174

7.55   (ai)   3.499   (aii)   2.998   (aiii)   2.365   (aiv)   1.895   (av)   1.415
(bi)   3.055   (bii)   2.681   (biii)   2.179   (biv)   1.782   (bv)   1.356
(ci)   2.787   (cii)   2.485   (ciii)   2.060   (civ)   1.708   (cv)   1.316

7.57   (ai) $H_0: \mu = 16;\ H_a: \mu > 16 \qquad \alpha = .10 \qquad$ Observed value: $t = \dfrac{18-16}{4/\sqrt{14}} = 1.87$
df = 13   Critical value: $t = 1.350$ Decision: Reject H_0 .025 < P-value < .05
(aii)   Type I
(bi) $H_0: \mu = 27;\ H_a: \mu < 27 \qquad \alpha = .05$
Observed value: $t = \dfrac{23-27}{7/\sqrt{9}} = -1.71$ df = 8   Critical value: $t = -1.860$
Decision: Do not reject H_0 .05 < P-value < .10
(bii)   Type II
(ci) $H_0: \mu = 30;\ H_a: \mu \neq 30 \qquad \alpha = .01$
Observed value: $t = \dfrac{25-30}{4/\sqrt{6}} = -3.06$ df = 5   Critical values: $t = \pm 4.032$
Decision: Do not reject H_0 .02 < P-value < .05
(cii)   Type II

(di) $H_0: \mu = 125; H_a: \mu > 125$ $\alpha = .05$

Observed value: $z = \dfrac{128 - 125}{18 / \sqrt{40}} = 1.05$ Critical value: $z = 1.65$

Decision: Do not reject H_0 P-value $= .5 - .3531 = .1469$

(dii) Type II

(ei) $H_0: \mu = 50; H_a: \mu < 50$ $\alpha = .025$

Observed value: $t = \dfrac{45 - 50}{10 / \sqrt{20}} = -2.24$ df $= 19$ Critical value: $t = -2.093$

Decision: Reject H_0 $.01 < P$-value $< .025$

(eii) Type I

(fi) $H_0: \mu = 60; H_a: \mu \neq 60$ $\alpha = .10$

Observed value: $t = \dfrac{70 - 60}{16 / \sqrt{8}} = 1.77$ df $= 7$ Critical values: $t = \pm 1.895$

Decision: Do not reject H_0 $.10 < P$-value $< .20$

(fii) Type II

7.59   (a) $.025 < P$-value $< .05$ P-value $< \alpha = .10$ Reject H_0
        (b) $.05 < P$-value $< .10$ P-value $> \alpha = .05$ Do not reject H_0
        (c) $.02 < P$-value $< .05$ P-value $> \alpha = .01$ Do not reject H_0
        (d) P-value $= .1469$ P-value $> \alpha = .05$ Do not reject H_0
        (e) $.01 < P$-value $< .025$ P-value $< \alpha = .025$ Reject H_0
        (f) $.10 < P$-value $< .20$ P-value $> \alpha = .10$ Do not reject H_0

7.61   (a) $H_0: \mu = 38{,}000; H_a: \mu > 38{,}000$ Observed value: $t = \dfrac{38{,}350 - 38{,}000}{420 / \sqrt{9}} = 2.50$

df $= 8$ Critical value: $t = 1.860$ for $\alpha = .05$, and $t = 2.896$ for $\alpha = .01$
Decision: Reject H_0 at the 5% level   Do not reject H_0 at the 1% level

        (b) $.01 < P$-value $< .025$ Reject H_0 at the 5% level, as P-value $< \alpha = .05$
Do not reject H_0 at the 1% level, as P-value $> \alpha = .01$

7.63 $H_0: \mu = 36; H_a: \mu \neq 36$ $\alpha = .05$ Observed value: $t = \dfrac{31.68 - 36}{9.27 / \sqrt{27}} = -2.42$ df $= 26$

Critical values: $t = \pm 2.056$ Decision: Reject H_0 $.02 < P$-value $< .05$
The data suggest that the mean trait anxiety level is not 36.

7.65 $H_0: \mu = 4.75; H_a: \mu < 4.75$ $\alpha = .01$ Observed value: $t = \dfrac{4.58 - 4.75}{.25 / \sqrt{10}} = -2.15$ df $= 9$

Critical value: $t = -2.821$ Decision: Do not reject H_0 $.025 < P$-value $< .05$
The data do not suggest that the mean acceleration is less than 4.75.

7.67   (a)   Yes

        (b) $\bar{x} = 1552/23 = 67.48$ $s = \sqrt{\dfrac{23(107{,}792) - (1552)^2}{(23)(22)}} = 11.80$

$H_0: \mu = 72; H_a: \mu \neq 72$ $\alpha = .05$ Observed value: $t = \dfrac{67.48 - 72}{11.80 / \sqrt{23}} = -1.84$

df $= 22$ Critical values: $t = \pm 2.074$ Decision: Do not reject H_0 $.05 < P$-value $< .10$
The data do not suggest that the mean test score is different from 72.

7.69 (a) Yes

(b) $H_0: \mu = 70$; $H_a: \mu > 70$ $\alpha = .05$ Observed value: $t = \dfrac{76.72 - 70}{10.99 / \sqrt{18}} = 2.59$ df = 17

Critical value: $t = 1.740$ Decision: Reject H_0 $.005 < P\text{-value} < .01$ The data suggest that the mean final exam score is statistically significantly larger than 70.

7.71 (a) Yes

(b) $H_0: \mu = 160$; $H_a: \mu < 160$ Observed value: $t = \dfrac{153.36 - 160}{9.47 / \sqrt{11}} = -2.33$ df = 10

$.01 < P\text{-value} < .025$ Reject the null hypothesis for any $\alpha \ge .025$

7.73 $\bar{x} = 226/10 = 22.6$ $s = \sqrt{\dfrac{10(5146) - (226)^2}{10(9)}}$ $H_0: \mu = 20$; $H_a: \mu > 20$ $\alpha = .01$

Observed value: $t = \dfrac{22.6 - 20}{\sqrt{\dfrac{10(5146) - (226)^2}{10(9)}} / \sqrt{10}} = 3.98$ df = 9 Critical value: $t = 2.821$

Decision: Reject H_0 $P\text{-value} < .005$
The data suggest that the mean time is more than 20 years.

7.75 (a) $28.7\% \pm (2.052)(5.7\%/\sqrt{28})$ or $26.49\% < \mu < 30.91\%$
(b) Yes, as the upper limit of the confidence interval is smaller than 36.2
(c) $E = (2.052)(5.7/\sqrt{28}) = 2.21$

7.77 (a) $4.67 \pm (2.145)(1.74/\sqrt{15})$ or $3.71 < \mu < 5.63$ (b) $(1/2)(10) = 5$

7.79 $16.5 \pm (2.074)(4.3/\sqrt{23})$ or $14.64 < \mu < 18.36$
The result does not suggest that μ is different from 15 minutes, because the confidence interval contains the value 15.

7.81 (a) $H_0: \mu = 78$; $H_a: \mu > 78$ Observed value: $t = 1.824$ df = 19
$\alpha = .05$ Critical value: $t = 1.729$ Decision: Reject H_0
$\alpha = .01$ Critical value: $t = 2.539$ Decision: Do not reject H_0
Note that the P-value is .042
(b) $80.08840 \pm (2.093)(5.12135/\sqrt{20})$ or $77.692 < \mu < 82.485$

7.83 (a) $\hat{p} = 200/400 = .50$ $.50 \pm 1.96\sqrt{\dfrac{(.50)(.50)}{400}}$ or $.45 < p < .55$

(b) $x = (.25)(900) = 225$ $.25 \pm 1.65\sqrt{\dfrac{(.25)(.75)}{900}}$ or $.23 < p < .27$

(c) $\hat{p} = 160/225 = .71$ $.71 \pm 2.58\sqrt{\dfrac{(.71)(.29)}{225}}$ or $.63 < p < .79$

(d) $x = (.82)(1250) = 1025$ $.82 \pm 1.96\sqrt{\dfrac{(.82)(.18)}{1250}}$ or $.80 < p < .84$

(e) $\hat{p} = 592/1600 = .37$ $.37 \pm 1.65\sqrt{\dfrac{(.37)(.63)}{1600}}$ or $.35 < p < .39$

(f) $x = (.61)(1000) = 610$ $.61 \pm 1.28\sqrt{\dfrac{(.61)(.39)}{1000}}$ or $.59 < p < .63$

7.85 $\hat{p} = 271/298 = .91$ $.91 \pm 1.96\sqrt{\dfrac{(.91)(.09)}{298}}$ or $.88 < p < .94$

7.87 $\hat{p} = 711/935 = .76$ $.76 \pm 1.96\sqrt{\dfrac{(.76)(.24)}{935}}$ or $.733 < p < .787$

7.89 $\hat{p} = 383/935 = .41$ $.41 \pm 1.96\sqrt{\dfrac{(.41)(.59)}{935}}$ or $.378 < p < .442$

7.91 $\dfrac{109}{223} \pm 1.96\sqrt{\dfrac{\left(\dfrac{109}{223}\right)\left(\dfrac{114}{223}\right)}{223}}$ or $.423 < p < .554$

7.93 (a) $\hat{p} = 45/75 = .60$

(b) $.60 \pm 1.96\sqrt{\dfrac{(.60)(.40)}{75}}$ or $.49 < p < .71$

(c) $E = 1.96\sqrt{\dfrac{(.60)(.40)}{75}} = .11$

7.95 (a) $\hat{p} = 260/500 = .52$ $.52 \pm 1.96\sqrt{\dfrac{(.52)(.48)}{500}}$ or $.48 < p < .56$

(b) $E = 1.96\sqrt{\dfrac{(.52)(.48)}{500}} = .04$

(c) No, as the confidence interval contains .50

7.97 $n = \left[\dfrac{1.96}{.04}\right]^2 \cdot \dfrac{1}{4} = 600.25$
The sample size should be at least 601

7.99 $n = \left[\dfrac{1.65}{.03}\right]^2 \cdot \dfrac{1}{4} = 756.25$
The sample size should be at least 757

7.101 (ai) $H_0: p = .53$; $H_a: p \neq .53$ (aii)   Two tailed
(aiii) Type I   The data suggest that p is not .53, when it is.
Type II   The data do not suggest that p is different from .53, when in fact it is different.
(bi) $H_0: p = .06$; $H_a: p < .06$ (bii)   Left tailed
(biii) Type I   The data suggest that p is less than .06, when it is not.
Type II   The data do not suggest that p is less than .06, when it is.
(ci) $H_0: p = .90$; $H_a: p < .90$ (cii)   Left tailed

(ciii) Type I   The data suggest that p is less than .90, when it is not.
Type II   The data do not suggest that p is less than .90, when it is.
(di) $H_0: p = .07; H_a: p > .07$ (dii)   Right tailed
(diii) Type I   The data suggest that p is more than .07, when it is not.
Type II   The data do not suggest that p is more than .07, when it is.
(ei) $H_0: p = .40; H_a: p \neq .40$ (eii)   Two tailed
(eiii) Type I   The data suggest that p is not .40, when it is.
Type II   The data do not suggest that p is different from .40, when in fact it is different.
(fi) $H_0: p = .05; H_a: p < .05$ (fii)   Left tailed
(fiii) Type I   The data suggest that p is less than .05, when it is not.
Type II   The data do not suggest that p is less than .05, when it is.
(gi) $H_0: p = .30; H_a: p > .30$ (gii)   Right tailed
(giii) Type I   The data suggest that p is more than .30, when it is not.
Type II   The data do not suggest that p is more than .30, when it is.

7.103 (a) $H_0: p = .8; H_a: \text{p} < .8$ Critical region: $z \leq -1.28$
       (b) $H_0: p = .8; H_a: \text{p} \neq .8$ Critical region: $z \leq -1.96$ or $z \geq 1.96$
       (c) $H_0: p = .8; H_a: \text{p} > .8$ Critical region: $z \geq 2.33$

7.105 (a) $H_0: p = .50; H_a: p > .50$ $\alpha = .05$ $\hat{p} = 660/1200 = .55$

$$\text{Observed value: } z = \frac{.55 - .50}{\sqrt{(.50)(.50)/1200}} = 3.46 \qquad \text{Critical value: } z = 1.65$$

Decision: Reject H_0
There is strong evidence suggesting that the true proportion is more than 50%.
       (b) As P-value $= P(z \geq 3.46) = .5 - .4997 = .0003 < \alpha = .05$, reject H_0

7.107 (a) $H_0: p = .75; H_a: p \neq .75$ $\alpha = .01$ $\hat{p} = 63/100 = .63$

$$\text{Observed value: } z = \frac{.63 - .75}{\sqrt{(.75)(.25)/100}} = -2.77 \qquad \text{Critical values: } z = \pm 2.58$$

Decision: Reject H_0 The data suggest that the proportion of subscribers renewing subscriptions is different from .75.
       (b) P-value $= 2P(z \leq -2.77) = 2(.5 - .4972) = .0056$
          As $.0056 < \alpha = .01$, then reject H_0

7.109 $H_0: p = .057; H_a: p < .057$ $\alpha = .025$ $\hat{p} = 15/405$

$$\text{Observed value: } z = \frac{\frac{15}{405} - .057}{\sqrt{(.057)(.943)/405}} = -1.73 \qquad \text{Critical value: } z = -1.96$$

Decision: Do not reject H_0 P-value $= P(z \leq -1.73) = .5 - .4582 = .0418$
The data do not suggest that the proportion p is less than .057.

7.111 (a) $H_0: p = .26; H_a: p < .26$ $\alpha = .05$ $\hat{p} = 54/225 = .24$

$$\text{Observed value: } z = \frac{.24 - .26}{\sqrt{(.26)(.74)/225}} = -.68 \qquad \text{Critical value: } z = -1.65$$

Decision: Do not reject H_0 P-value $= P(z \leq -.68) = (.5 - .2518) = .2482$
At the 5% level, the data do not support the administrator's claim that fewer than 26% of children live with a single parent.

(b) $.24 \pm 1.65\sqrt{\dfrac{(.24)(.76)}{225}}$ or $.19 < p < .29$ The result agrees with the conclusion in part (a), as the confidence interval contains the value .26.

7.113 (a) H_0: $p = .40$; H_a: $p > .40$ $\alpha = .05$ $\hat{p} = 138/300 = .46$

Observed value: $z = \dfrac{.46 - .40}{\sqrt{(.40)(.60)/300}} = 2.12$ Critical value: $z = 1.65$

Decision: Reject H_0 P-value $= P(z \geq 2.12) = (.5 - .4830) = .0170$
At the 5% level, the data support the administrator's claim that more than 40% of children live in homes where both parents work to make ends meet.

(b) $.46 \pm 1.65\sqrt{\dfrac{(.46)(.54)}{300}}$ or $.41 < p < .51$ The result agrees with the conclusion in part (a), as the confidence interval does not contain the value .40.

7.115 $29,400 \pm (2.58)(6325/\sqrt{40})$ or $26,819.82 < \mu < 31,980.18$
Yes, as 26,819.82 is larger than 25,100

7.117 (a) $\bar{x} = 2624/40 = 65.60$ $s = \sqrt{\dfrac{40(173,804) - (2624)^2}{40(39)}} = 6.54$

$65.60 \pm (1.96)(6.54/\sqrt{40})$ or $63.57 < \mu < 67.63$
(b) Yes, as 63.57 is larger than 62.

7.119 $n = \left[\dfrac{(1.96)(17)}{2}\right]^2 = 277.56$ The sample size should be at least 278

7.121 (a) H_0: $\mu = 450$; H_a: $\mu \neq 450$ $\alpha = .01$ Observed value: $z = \dfrac{450.4 - 450}{.95/\sqrt{43}} = 2.76$
Critical values: $z = \pm 2.58$ Decision: Reject H_0
The data suggest that the mean length is not 450 inches.
(b) P-value $= 2P(z \geq 2.76) = 2(.5 - .4971) = .0058$
Reject the null hypothesis as $.0058 < \alpha = .01$

7.123 (a) H_0: $\mu = 35$; H_a: $\mu > 35$ $\alpha = .05$ Observed value: $z = \dfrac{38 - 35}{7/\sqrt{40}} = 2.71$
Critical value: $z = 1.65$ Decision: Reject H_0
The data suggest that the level is more than 35.
(b) P-value $= P(z \geq 2.71) = .5 - .4966 = .0034$
Reject the null hypothesis as $.0034 < \alpha = .05$

7.125 H_0: $\mu = 12$; H_a: $\mu < 12$ $\alpha = .05$ Observed value: $z = \dfrac{11.85 - 12}{.65/\sqrt{38}} = -1.42$
Critical value: $z = -1.65$ Decision: Do not reject H_0
P-value $= P(z \leq -1.42) = .5 - .4222 = .0778$
The data do not suggest that the mean length is less than 12 inches.

7.127 (a) 2.718 (b) 1.714 (c) 1.345 (d) −2.080

7.129 (a) $H_0: \mu = 2200$; $H_a: \mu > 2200$ $\quad \alpha = .05$ $\quad$ Observed value: $t = \dfrac{2315 - 2200}{115/\sqrt{20}} = 4.47$

$\quad\quad$ df = 19 $\quad$ Critical value: $t = 1.729$ $\quad$ Decision: Reject H_0
$\quad\quad$ The data suggest that the mean number of passengers per day is more than 2200.
$\quad$ (b) P-value < .005 $\quad\quad$ Reject H_0, as P-value < $\alpha = .05$

7.131 $H_0: \mu = 13$; $H_a: \mu \neq 13$ $\quad \alpha = .05$ $\quad$ Observed value: $t = \dfrac{10.46 - 13}{5.57/\sqrt{27}} = -2.37$

$\quad$ df = 26 $\quad$ Critical values: $t = \pm 2.056$ $\quad$ Decision: Reject H_0 $\quad$ .02 < P-value < .05
$\quad$ The data suggest that the mean tension score is not 13.

7.133 (a) Yes

$\quad$ (b) $H_0: \mu = 600$; $H_a: \mu < 600$ $\quad \alpha = .05$ $\quad$ Observed value: $t = \dfrac{517.22 - 600}{110.87/\sqrt{18}} = -3.17$

$\quad\quad$ df = 17 $\quad$ Critical value: $t = -1.740$ $\quad$ Decision: Reject H_0
$\quad\quad$ The data suggest that the mean tax deduction for charities is less than \$600.
$\quad\quad$ P-value < .005 $\quad\quad$ Reject H_0, as P-value < .05

7.135 (a) $517.22 \pm (1.740)(110.87/\sqrt{18})$ or $471.75 < \mu < 562.69$
$\quad$ (b) $E = (1.740)(110.87/\sqrt{18}) = 45.47$

7.137 (a) $x = (.50)(100) = 50$ $\quad\quad$ $.50 \pm 1.96\sqrt{\dfrac{(.50)(.50)}{100}}$ or $.40 < p < .60$

$\quad$ (b) $\hat{p} = 50/125 = .40$ $\quad\quad$ $.40 \pm 1.65\sqrt{\dfrac{(.40)(.60)}{125}}$ or $.33 < p < .47$

$\quad$ (c) $x = (.72)(400) = 288$ $\quad\quad$ $.72 \pm 2.58\sqrt{\dfrac{(.72)(.28)}{400}}$ or $.66 < p < .78$

$\quad$ (d) $\hat{p} = 45/225 = .20$ $\quad\quad$ $.20 \pm 1.96\sqrt{\dfrac{(.20)(.80)}{225}}$ or $.15 < p < .25$

$\quad$ (e) $x = (.60)(60) = 36$ $\quad\quad$ $.60 \pm 1.65\sqrt{\dfrac{(.60)(.40)}{60}}$ or $.50 < p < .70$

$\quad$ (f) $\hat{p} = 30/50 = .60$ $\quad\quad$ $.60 \pm 2.58\sqrt{\dfrac{(.60)(.40)}{50}}$ or $.42 < p < .78$

7.139 $.54 \pm 1.96\sqrt{\dfrac{(.54)(.46)}{499}}$ or $.50 < p < .58$

7.141 $.37 \pm 1.65\sqrt{\dfrac{(.37)(.63)}{499}}$ or $.33 < p < .41$

7.143 (a) $H_0: p = .40; H_a: p \neq .40$ $\alpha = .05$ $\hat{p} = .37$

Observed value: $z = \dfrac{.37 - .40}{\sqrt{(.40)(.60)/1200}} = -2.12$ Critical values: $z = \pm 1.65$

Decision: Reject H_0 P-value $= 2P(z \leq -2.12) = 2(.5 - .4830) = .0340$

At the 5% level, the data indicate that the percentage of working women with employer provided health insurance is different from 40%.

(b) No, as the P-value is larger than .01

Chapter 8

8.1 (ai) $H_0: \mu = 0; H_a: \mu \neq 0$ $\alpha = .05$ Observed value: $t = \dfrac{1.63 - 0}{2.83/\sqrt{8}} = 1.63$

df $= 8 - 1 = 7$ Critical values: $t = \pm 2.365$ Decision: Do not reject H_0
$.10 < P$-value $< .20$

(aii) Type II

(bi) $H_0: \mu = 0; H_a: \mu < 0$ $\alpha = .01$ Observed value: $t = \dfrac{-1.86 - 0}{3.72/\sqrt{7}} = -1.32$

df $= 7 - 1 = 6$ Critical value: $t = -3.143$ Decision: Do not reject H_0
$.10 < P$-value $< .25$

(bii) Type II

(ci) $H_0: \mu = 0; H_a: \mu > 0$ $\alpha = .05$ The set of x values is $\{-2, 5, 6, 3, 3, 8\}$

$\Sigma x = 23$ $\Sigma x^2 = 147$ $\bar{x} = 23/6$ $s = \sqrt{\dfrac{6(147) - (23)^2}{(6)(5)}}$

Observed value: $t = \dfrac{23/6 - 0}{\sqrt{\dfrac{6(147) - (23)^2}{(6)(5)}}/\sqrt{6}} = 2.74$ df $= 6 - 1 = 5$

Critical value: $t = 2.015$ Decision: Reject H_0 $.01 < P$-value $< .025$

(cii) Type I

8.3 $H_0: \mu = 0; H_a: \mu \neq 0$ $\alpha = .05$
The set of x values is $\{8, 12, 6, -2, -14, 16, 6, 12, -2, 8, 8, 14\}$

Observed value: $t = \dfrac{6 - 0}{8.40/\sqrt{12}} = 2.47$ df $= 12 - 1 = 11$ Critical values: $t = \pm 2.201$

Decision: Reject H_0 $.02 < P$-value $< .05$
The data suggest that the mean systolic blood pressure is different between smokers and nonsmokers.

8.5 (a) $H_0: \mu = 0; H_a: \mu < 0$ Observed value: $t = \dfrac{-.225 - 0}{.276/\sqrt{8}} = -2.31$ df $= 8 - 1 = 7$

$.025 < P$-value $< .05$

(b) Reject H_0 at the 5% and 10% levels, but do not reject H_0 at the 1% level of significance.

8.7 $H_0: \mu = 0$; $H_a: \mu > 0$ $\quad \alpha = .10$ $\quad$ The set of x values is $\{3, 5, -2, 2, -1, 3\}$

$\sum x = 10$ $\quad \sum x^2 = 52$ $\quad \bar{x} = 10/6$ $\quad s = \sqrt{\dfrac{6(52) - (10)^2}{(6)(5)}}$

Observed value: $t = \dfrac{10/6 - 0}{\sqrt{\dfrac{6(52) - (10)^2}{(6)(5)}} / \sqrt{6}} = 1.54$ $\quad$ df $= 6 - 1 = 5$

Critical value: $t = 1.476$ $\quad$ Decision: Reject H_0 $\quad$ $.05 < P$-value $< .10$
The data suggest that the advertising program is effective.

8.9 (a) $1.63 \pm (2.365)(2.83/\sqrt{8})$ or $-.74 < \mu < 4.00$
(b) $-1.86 \pm (3.143)(3.72/\sqrt{7})$ or $-6.28 < \mu < 2.56$

(c) $23/6 \pm (2.015)\left(\sqrt{\dfrac{6(147) - 23^2}{(6)(5)}} / \sqrt{6} \right)$ or $1.01 < \mu < 6.65$

8.11 (a) $6 \pm (2.201)(8.40/\sqrt{12})$ or $.66 < \mu < 11.34$
(b) $-.225 \pm (1.895)(.276/\sqrt{8})$ or $-.41 < \mu < -.04$

(c) $10/6 \pm (1.476)\left(\sqrt{\dfrac{6(52) - (10)^2}{(6)(5)}} / \sqrt{6} \right)$ or $.065 < \mu < 3.270$

8.13 (a) $H_0: \mu_A - \mu_B = 0$; $H_a: \mu_A - \mu_B \neq 0$ $\quad \alpha = .05$

Observed value: $z = \dfrac{(175 - 165) - 0}{\sqrt{\dfrac{360}{40} + \dfrac{350}{50}}} = 2.50$ $\quad$ Critical values: $z = \pm 1.96$

Decision: Reject H_0 $\quad$ P-value $= 2P(z \geq 2.50) = 2(.0062) = .0124$
(b) $H_0: \mu_A - \mu_B = 0$; $H_a: \mu_A - \mu_B > 0$ $\quad \alpha = .05$

Observed value: $z = \dfrac{(400 - 396) - 0}{\sqrt{\dfrac{210}{35} + \dfrac{105}{35}}} = 1.33$ $\quad$ Critical value: $z = 1.65$

Decision: Do not reject H_0 $\quad$ P-value $= P(z \geq 1.33) = .0918$
(c) $H_0: \mu_A - \mu_B = 0$; $H_a: \mu_A - \mu_B < 0$ $\quad \alpha = .05$

Observed value: $z = \dfrac{(42 - 48) - 0}{\sqrt{\dfrac{260}{52} + \dfrac{216}{54}}} = -2$ $\quad$ Critical value: $z = -1.65$

Decision: Reject H_0 $\quad$ P-value $= P(z \leq -2) = .0228$

8.15 (a) $H_0: \mu_A - \mu_B = 0$; $H_a: \mu_A - \mu_B \neq 0$ $\quad \alpha = .05$

Observed value: $z = \dfrac{(65.39 - 64.12) - 0}{\sqrt{\dfrac{7.02}{57} + \dfrac{4.75}{66}}} = 2.88$

Critical value: $z = \pm 1.96$ $\quad$ Decision: Reject H_0
The data suggest that the mean height of soprano and alto singers, on average, are different.

(b) P-value $= 2P(z \geq 2.88) = 2(.0020) = .0040$
As $.0040 < .05$, reject H_0 at the 5% level.

8.17 (a) $H_0: \mu_A - \mu_B = 0$; $H_a: \mu_A - \mu_B > 0$ $\alpha = .01$

Observed value: $z = \dfrac{(70.99 - 69.41) - 0}{\sqrt{\dfrac{(2.52)^2}{65} + \dfrac{(2.79)^2}{42}}} = 2.97$ Critical value: $z = 2.33$

Decision: Reject H_0
The data suggest that, on average, bass singers are taller than tenor singers.

(b) P-value $= P(z \geq 2.97) = .0015$
As $.0015 < .01$, reject the null hypothesis at the 1% level.

8.19 (a) $(65.39 - 64.12) \pm (1.96)\left(\sqrt{\dfrac{7.02}{57} + \dfrac{4.75}{66}}\right)$ or $.40 < \mu_A - \mu_B < 2.14$

(b) $(70.99 - 69.41) \pm (2.33)\left(\sqrt{\dfrac{(2.52)^2}{65} + \dfrac{(2.79)^2}{42}}\right)$ or $.34 < \mu_A - \mu_B < 2.82$

8.21 (a) $H_0: \mu_A - \mu_B = 0$; $H_a: \mu_A - \mu_B \neq 0$ $\alpha = .05$

Observed value: $t = \dfrac{(83 - 80) - 0}{\sqrt{\dfrac{21}{8} + \dfrac{5}{12}}} = 1.72$ df = smaller of $(8 - 1, 12 - 1) = 7$

Critical values: $t = \pm 2.365$ Decision: Do not reject H_0 $.10 < P$-value $< .20$
The data do not suggest that the population means are different.

(b) $H_0: \mu_A - \mu_B = 0$; $H_a: \mu_A - \mu_B > 0$ $\alpha = .01$

Observed value: $t = \dfrac{(46 - 41) - 0}{\sqrt{\dfrac{39}{15} + \dfrac{17}{13}}} = 2.53$ df = smaller of $(15 - 1, 13 - 1) = 12$

Critical value: $t = 2.681$ Decision: Do not reject H_0 $.01 < P$-value $< .025$
The data do not suggest that the mean of population A is less than the mean of
population B

(c) $H_0: \mu_A - \mu_B = 0$; $H_a: \mu_A - \mu_B < 0$ $\alpha = .10$

Observed value: $t = \dfrac{(180 - 200) - 0}{\sqrt{\dfrac{70}{10} + \dfrac{340}{7}}} = -2.68$ df = smaller of $(10 - 1, 7 - 1) = 6$

Critical value: $t = -1.440$ Decision: Reject H_0 $.01 < P$-value $< .025$
The data suggest that the mean of population A is less than the mean of population B.

8.23 (a) $H_0: \mu_A - \mu_B = 0$; $H_a: \mu_A - \mu_B < 0$ $\alpha = .05$

Observed value: $t = \dfrac{(63.71 - 72.13) - 0}{\sqrt{\dfrac{(8.16)^2}{7} + \dfrac{(4.84)^2}{16}}} = -2.54$

df = smaller of $(7 - 1, 16 - 1) = 6$ Critical value: $t = -1.943$
Decision: Reject H_0
The data suggest that the mean launch temperature for flights with O–ring failure is less than for flights with no O–ring failure.

(b) $(63.71 - 72.13) \pm (2.447)\sqrt{\dfrac{(8.16)^2}{7} + \dfrac{(4.84)^2}{16}}$ or $-16.53 < \mu_A - \mu_B < -.31$

8.25 (a) Yes

(b) $H_0: \mu_A - \mu_B = 0$; $H_a: \mu_A - \mu_B \neq 0$ $\alpha = .05$

Observed value: $t = \dfrac{(76.72 - 70.86) - 0}{\sqrt{\dfrac{(10.99)^2}{18} + \dfrac{(14.37)^2}{14}}} = 1.26$

df = smaller of $(18 - 1, 14 - 1) = 13$ Critical values: $t = \pm 2.160$
Decision: Do not reject H_0 $.20 < P\text{-value} < .50$
The data do not suggest that there is a difference in mean final exam scores between the two sections.

8.27 (a) $H_0: \mu_E - \mu_A = 0$; $H_a: \mu_E - \mu_A > 0$ $\alpha = .05$

Observed value: $t = \dfrac{(65.37 - 62.59) - 0}{\sqrt{\dfrac{(8.38)^2}{16} + \dfrac{(5.59)^2}{17}}} = 1.11$

df = smaller of $(16 - 1, 17 - 1) = 15$ Critical value: $t = 1.753$
Decision: Do not reject H_0 $.10 < P\text{-value} < .25$
The data do not suggest that the population means are different.

(b) Based on the P-value $= .142$, we would expect about $(.142)(1000) = 142$ mean differences as large or larger than 2.78. So the value of 138 is consistent with the result in part (a).

8.29 (a) $(83 - 80) \pm (2.365)\sqrt{\dfrac{21}{8} + \dfrac{5}{12}}$ or $-1.12 < \mu_A - \mu_B < 7.12$

(b) $(46 - 41) \pm (3.005)\sqrt{\dfrac{39}{15} + \dfrac{17}{13}}$ or $-.94 < \mu_A - \mu_B < 10.94$

(c) $(180 - 200) \pm (1.440)\sqrt{\dfrac{70}{10} + \dfrac{340}{7}}$ or $-30.73 < \mu_A - \mu_B < -9.27$

8.31 (a) $H_0: \mu_1 - \mu_2 = 0$; $H_a: \mu_1 - \mu_2 > 0$ $\alpha = .05$ Observed value: $t = 3.441$
df = 27   Critical value: $t = 1.703$ Decision: Reject H_0 P-value $< .005$
The data support the engineer's suspicion that the power plant was substantially increasing the air pollution in the vicinity of the plant.

(b) $5.33 < \mu_1 - \mu_2 < 21.1$

8.33 $H_0: \mu_A - \mu_B = 0;\ H_a: \mu_A - \mu_B < 0$ $\alpha = .05$

Observed value: $t = \dfrac{(120-125)-0}{\sqrt{\dfrac{5(100)+9(81)}{6+10-2}}\sqrt{\dfrac{1}{6}+\dfrac{1}{10}}} = -1.03$

df $= 6 + 10 - 2 = 14$ Critical values: $t = -1.761$ Decision: Do not reject H_0
$.10 < P\text{-value} < .25$ The data do not suggest that the mean of population A is less than the mean of population B.

8.35   (a)   Let A = women and B = men $H_0: \mu_A - \mu_B = 30;\ H_a: \mu_A - \mu_B \neq 30$ $\alpha = .10$

Observed value: $t = \dfrac{(206.16-182.55)-30}{\sqrt{\dfrac{14(17.69)^2+14(17.66)^2}{15+15-2}}\sqrt{\dfrac{1}{15}+\dfrac{1}{15}}} = -.99$

df $= 15 + 15 - 2 = 28$ Critical values: $t = \pm 1.701$
Decision: Do not reject H_0 $.20 < P\text{-value} < .50$ The data do not suggest that the difference in the population mean times is not equal to 30 minutes.

      (b) $(206.16 - 182.55) \pm (1.701)\left(\sqrt{\dfrac{14(17.66)^2+14(17.69)^2}{15+15-2}}\sqrt{\dfrac{1}{15}+\dfrac{1}{15}} \right)$ or

$12.63 < \mu_A - \mu_B < 34.59$ Note that not rejecting H_0 at the 10% level corresponds to the 90% confidence interval containing the hypothesized value 30.

8.37   (a) $H_0: p_1 - p_2 = 0;\ H_a: p_1 - p_2 \neq 0$ $\alpha = .05$

$\hat{p}_1 = 175/250$ $\hat{p}_2 = 135/175$ $\hat{p} = \dfrac{175+135}{250+175} = 310/425$

Observed value: $z = \dfrac{\left(\dfrac{175}{250} - \dfrac{135}{175}\right)-0}{\sqrt{\left(\dfrac{310}{425}\right)\left(\dfrac{115}{425}\right)\left[\dfrac{1}{250}+\dfrac{1}{175}\right]}} = -1.63$

Critical values: $z = \pm 1.96$ Decision: Do not reject H_0
$P\text{-value} = 2P(z \leq -1.63) = 2(.0516) = .1032$
The data do not suggest that the population proportions p_1 and p_2 are different.

      (b) $H_0: p_1 - p_2 = 0;\ H_a: p_1 - p_2 > 0$ $\alpha = .10$

$\hat{p}_1 = 195/375$ $\hat{p}_2 = 150/325$ $\hat{p} = \dfrac{195+150}{375+325} = 345/700$

Observed value: $z = \dfrac{\left(\dfrac{195}{375} - \dfrac{150}{325}\right)-0}{\sqrt{\left(\dfrac{345}{700}\right)\left(\dfrac{355}{700}\right)\left[\dfrac{1}{375}+\dfrac{1}{325}\right]}} = 1.54$ Critical value: $z = 1.28$

Decision: Reject H_0 $P\text{-value} = P(z \geq 1.54) = .0618$ The data suggest that the population A proportion p_1 is larger than that of the population B proportion p_2.

      (c) $H_0: p_1 - p_2 = 0;\ H_a: p_1 - p_2 < 0$ $\alpha = .05$

$\hat{p}_1 = 175/425$ $\hat{p}_2 = 205/400$ $\hat{p} = \dfrac{175+205}{425+400} = 380/825$

Observed value: $z = \dfrac{\left(\dfrac{175}{425} - \dfrac{205}{400}\right) - 0}{\sqrt{\left(\dfrac{380}{825}\right)\left(\dfrac{445}{825}\right)\left[\dfrac{1}{425} + \dfrac{1}{400}\right]}} = -2.90$

Critical value: $z = -1.65$ Decision: Reject H_0

P-value $= P(z \le -2.90) = .0019$ The data suggest that the population A proportion p_1 is smaller than that of the population B proportion p_2.

8.39   (a) $\left(\dfrac{175}{250} - \dfrac{135}{175}\right) \pm (1.96)\sqrt{\left(\dfrac{175}{250}\right)\left(\dfrac{75}{250}\right)\left(\dfrac{1}{250}\right) + \left(\dfrac{135}{175}\right)\left(\dfrac{40}{175}\right)\left(\dfrac{1}{175}\right)}$ or

         $-.156 < p_1 - p_2 < .013$

       (b) $\left(\dfrac{195}{375} - \dfrac{150}{325}\right) \pm (1.28)\sqrt{\left(\dfrac{195}{375}\right)\left(\dfrac{180}{375}\right)\left(\dfrac{1}{375}\right) + \left(\dfrac{150}{325}\right)\left(\dfrac{175}{325}\right)\left(\dfrac{1}{325}\right)}$ or

         $.010 < p_1 - p_2 < .107$

       (c) $\left(\dfrac{175}{425} - \dfrac{205}{400}\right) \pm (1.65)\sqrt{\left(\dfrac{175}{425}\right)\left(\dfrac{250}{425}\right)\left(\dfrac{1}{425}\right) + \left(\dfrac{205}{400}\right)\left(\dfrac{195}{400}\right)\left(\dfrac{1}{400}\right)}$ or

         $-.158 < p_1 - p_2 < -.044$

8.41   (a) $H_0: p_B - p_W = 0;\ H_a: p_B - p_W < 0$ $\alpha = .025$

         $\hat{p}_B = 40/55$ $\hat{p}_W = 99/113$ $\hat{p} = \dfrac{40 + 99}{55 + 113} = 139/168$

         Observed value: $z = \dfrac{\left(\dfrac{40}{55} - \dfrac{99}{113}\right) - 0}{\sqrt{\left(\dfrac{139}{168}\right)\left(\dfrac{29}{168}\right)\left[\dfrac{1}{55} + \dfrac{1}{113}\right]}} = -2.40$ Critical value: $z = -1.96$

         Decision: Reject H_0 P-value $= P(z \le -2.40) = .0082$

       (b) $\hat{p}_B / \hat{p}_W = \left(\dfrac{40/55}{99/113}\right) = .83$

         The data do not meet the requirements of the 80% rule.

8.43   Let p_Z and p_N be the population proportions of smokers that quit smoking using Zyban and a nicotine patch, respectively.

       (a) $H_0: p_Z - p_N = 0;\ H_a: p_Z - p_N > 0$ $\alpha = .01$ $\hat{p}_Z = 109/223$ $\hat{p}_N = 80/223$

         $\hat{p} = \dfrac{109 + 80}{223 + 223} = 189/446$ Observed value: $z = \dfrac{\left(\dfrac{109}{223} - \dfrac{80}{223}\right) - 0}{\sqrt{\left(\dfrac{189}{446}\right)\left(\dfrac{257}{446}\right)\left[\dfrac{1}{223} + \dfrac{1}{223}\right]}} = 2.78$

         Critical value: $z = 2.33$ Decision: Reject H_0

       (b) P-value $= P(z \ge 2.78) = .0027 < .01$ Reject H_0

         The data suggest that the drug Zyban is more effective than the nicotine patch in helping smokers quit smoking.

8.45 Let p_A and p_B be the death rates for two populations of patients, one treated with atherectomy (p_A), the other with angioplasty (p_B).

$H_0: p_A - p_B = 0$; $H_a: p_A - p_B \neq 0$ Now $\dfrac{205}{5000} = .041$ is a simulated P-value.

As $.041 < .05$, there is evidence to suggest at the 5% level that there is a difference in the population proportions.

8.47 (a) $\left(\dfrac{40}{55} - \dfrac{99}{113}\right) \pm (1.96)\sqrt{\left(\dfrac{40}{55}\right)\left(\dfrac{15}{55}\right)\left(\dfrac{1}{55}\right) + \left(\dfrac{99}{113}\right)\left(\dfrac{14}{113}\right)\left(\dfrac{1}{113}\right)}$ or

$-.281 < p_B - p_W < -.016$

(b) $\left(\dfrac{109}{223} - \dfrac{80}{223}\right) \pm (2.58)\sqrt{\left(\dfrac{109}{223}\right)\left(\dfrac{114}{223}\right)\left(\dfrac{1}{223}\right) + \left(\dfrac{80}{223}\right)\left(\dfrac{143}{223}\right)\left(\dfrac{1}{223}\right)}$ or

$.010 < p_Z - p_N < .250$

8.49 The set of differences x is $\{-1, -2.4, 1, .3, -4.5, -2.6, -.3, -2.3, 1.1, -3.2\}$

$\Sigma x = -13.9$ $\Sigma x^2 = 51.69$ $\bar{x} = -13.9/10 = -1.39$

$s = \sqrt{\dfrac{(10)(51.69) - (-13.9)^2}{(10)(9)}} \doteq 1.896$

(a) $H_0: \mu = 0$; $H_a: \mu < 0$ $\alpha = .05$ Observed value: $t = \dfrac{-1.39 - 0}{1.896 / \sqrt{10}} = -2.32$

df $= 10 - 1 = 9$ Critical value: $t = -1.833$ Decision: Reject H_0

(b) $.01 < P$-value $< .025$ As P-value $< .05$, reject H_0

8.51 (a) $-1.39 \pm (1.833)(1.896/\sqrt{10})$ or $-2.489 < \mu_A - \mu_B < -.291$

It's quicker by car with a range of minutes between .3 and 2.5 minutes.

(b) $11/7 \pm (2.447)\left(\sqrt{\dfrac{7(91) - (11)^2}{(7)(6)}} / \sqrt{7}\right)$ or $-1.67 < \mu_A - \mu_B < 4.81$

You cannot conclude that one of the fertilizers is more effective, because the confidence interval contains the value 0.

8.53 $H_0: \mu_A - \mu_B = 0$; $H_a: \mu_A - \mu_B \neq 0$ $\alpha = .01$

Observed value: $z = \dfrac{(37.4 - 41.8) - 0}{\sqrt{\dfrac{(4.9)^2}{30} + \dfrac{(5.7)^2}{33}}} = -3.29$ Critical values: $z = \pm 2.58$

Decision: Reject H_0 P-value $= P(z \leq -3.29) = 2(.0005) = .0010$

The data suggest that the population means are different.

8.55 (a) $(42.70 - 43.17) \pm (1.96)\left(\sqrt{\dfrac{(7.68)^2}{40} + \dfrac{(6.75)^2}{40}}\right)$ or $-3.64 < \mu_A - \mu_N < 2.70$

(b) $H_0: \mu_A - \mu_N = 0$; $H_a: \mu_A - \mu_N \neq 0$ $\alpha = .05$

Observed value: $z = \dfrac{(42.70 - 43.17) - 0}{\sqrt{\dfrac{(7.68)^2}{40} + \dfrac{(6.75)^2}{40}}} = -.29$

Critical values: $z = \pm 1.96$ Decision: Do not reject H_0

The data do not suggest that the population means are different. This is consistent with part (a), as the 95% confidence interval contains the value 0.

8.57 H_0: $\mu_F - \mu_M = 5$; H_a: $\mu_F - \mu_M \neq 5$ $\alpha = .02$

Observed value: $t = \dfrac{(48.1 - 39.6) - 5}{\sqrt{\dfrac{(6.8)^2}{10} + \dfrac{(4.4)^2}{10}}} = 1.37$ df = smaller of $(10 - 1, 10 - 1) = 9$

Critical values: $t = \pm 2.821$ Decision: Do not reject H_0 $.20 < P\text{-value} < .50$

The data do not suggest that the population mean female six site fold is 5 millimeters more than the population mean for males.

8.59   (a) $(4.8 - 6.9) \pm (1.753)\left(\sqrt{\dfrac{4.3}{18} + \dfrac{12.7}{16}}\right)$ or $-3.88 < \mu_A - \mu_B < -.32$ You can

conclude that $\mu_A < \mu_B$, because the confidence interval does not contain the value 0.

      (b) $(9.1 - 9.9) \pm (2.602)\left(\sqrt{\dfrac{1.05}{16} + \dfrac{3.63}{16}}\right)$ or $-2.21 < \mu_A - \mu_B < .61$ You cannot

conclude that $\mu_A < \mu_B$, because the confidence interval contains the value 0.

8.61 H_0: $p_S - p_F = 0$; H_a: $p_S - p_F \neq 0$ $\alpha = .05$

$\hat{p}_S = 70/200$ $\hat{p}_F = 45/100$ $\hat{p} = \dfrac{70 + 45}{200 + 100} = 115/300$

Observed value: $z = \dfrac{\left(\dfrac{70}{200} - \dfrac{45}{100}\right) - 0}{\sqrt{\left(\dfrac{115}{300}\right)\left(\dfrac{185}{300}\right)\left[\dfrac{1}{200} + \dfrac{1}{100}\right]}} = -1.68$ Critical values: $z = \pm 1.96$

Decision: Do not reject H_0 $P\text{-value} = 2P(z \leq -1.68) = 2(.0465) = .0930$

8.63   (a) H_0: $p_A - p_B = 0$; H_a: $p_A - p_B \neq 0$ $\alpha = .01$

         $\hat{p}_A = 70/125$ $\hat{p}_B = 130/200$ $\hat{p} = \dfrac{70 + 130}{125 + 200} = 200/325$

         Observed value: $z = \dfrac{\left(\dfrac{70}{125} - \dfrac{130}{200}\right) - 0}{\sqrt{\left(\dfrac{200}{325}\right)\left(\dfrac{125}{325}\right)\left[\dfrac{1}{125} + \dfrac{1}{200}\right]}} = -1.62$

         Critical values: $z = \pm 2.58$ Decision: Do not reject H_0

      (b) $P\text{-value} = 2P(z \leq -1.62) = 2(.0526) = .1052 > .01$

Do not reject the null hypothesis. The data do not suggest a difference in population proportions.

8.65   (a) $\left(\dfrac{70}{200} - \dfrac{45}{100}\right) \pm (1.96)\sqrt{\left(\dfrac{70}{200}\right)\left(\dfrac{130}{200}\right)\left(\dfrac{1}{200}\right) + \left(\dfrac{45}{100}\right)\left(\dfrac{55}{100}\right)\left(\dfrac{1}{100}\right)}$ or

$-.218 < p_A - p_B < .018$

(b) $\left(\dfrac{70}{125}-\dfrac{130}{200}\right) \pm (2.58)\sqrt{\left(\dfrac{70}{125}\right)\left(\dfrac{55}{125}\right)\left(\dfrac{1}{125}\right)+\left(\dfrac{130}{200}\right)\left(\dfrac{70}{200}\right)\left(\dfrac{1}{200}\right)}$ or

$-.234 < p_A - p_B < .054$

You cannot conclude that $p_A \neq p_B$, because the confidence intervals contain the value 0.

Chapter 9

9.1 $H_0: \beta_1 = 0; H_a: \beta_1 > 0$ $\qquad \alpha = .01$

x	y	x^2	y^2	xy
1	2	1	4	2
1	4	1	16	4
3	6	9	36	18
3	8	9	64	24
5	10	25	100	50
5	12	25	144	60
7	14	49	196	98
7	16	49	256	112
Sums: 32	72	168	816	368

$n = 8$

$$r = \dfrac{8(368) - (32)(72)}{\sqrt{8(168) - (32)^2}\sqrt{8(816) - (72)^2}} = .9759001$$

Observed value: $t = .9759001\sqrt{\dfrac{8-2}{1-(.9759001)^2}} = 10.95$ $\qquad$ df $= 8 - 2 = 6$

Critical value: $t = 3.143$ $\qquad$ Decision: Reject H_0 $\qquad$ P-value $< .005$

The data suggest that there is a positive linear relationship between the x and y values for the population.

9.3 $H_0: \beta_1 = 0; H_a: \beta_1 < 0$ $\qquad \alpha = .05$

x	y	x^2	y^2	xy
4	12	16	144	48
4	11	16	121	44
6	10	36	100	60
6	9	36	81	54
8	7	64	49	56
8	6	64	36	48
Sums: 36	55	232	531	310

$n = 6$

$$r = \dfrac{6(310) - (36)(55)}{\sqrt{6(232) - (36)^2}\sqrt{6(531) - (55)^2}} = -.9652342$$

Observed value: $t = -.9652342\sqrt{\dfrac{6-2}{1-(-.9652342)^2}} = -7.39$ $\qquad$ df $= 6 - 2 = 4$

Critical value: $t = -2.132$ $\qquad$ Decision: Reject H_0 $\qquad$ P-value $< .005$

The data suggest that there is a negative linear relationship between the x and y values for the population.

9.5 $H_0: \beta_1 = 0$; $H_a: \beta_1 > 0$ $\alpha = .05$ $r = \dfrac{5(232)-(15)(65)}{\sqrt{5(55)-(15)^2}\,\sqrt{5(983)-(65)^2}} = .996007$

Observed value: $t = r\sqrt{\dfrac{n-2}{1-r^2}} = (.996007)\sqrt{\dfrac{5-2}{1-(.996007)^2}} = 19.32$ df $= 5 - 2 = 3$

Critical value: $t = 2.353$ Decision: Reject H_0 P-value $< .005$ The data suggest that there is a linear relationship between sales and intensity of advertising.

9.7 Observed value: $t = -12.6$
The test is statistically significant at the 1% level, as P-value $\le (1/2)(.0002) < .01$.

9.9 $H_0: \beta_1 = 0$; $H_a: \beta_1 \neq 0$ $\alpha = .05$ $n = 11$

Observed value: $t = r\sqrt{\dfrac{n-2}{1-r^2}} = (.844416)\sqrt{\dfrac{11-2}{1-(.844416)^2}} = 4.729$ df $= 11 - 2 = 9$

Critical values: $t = \pm 2.262$ Decision: Reject H_0 P-value $< 2(.005) = .01$ The data suggest that there is a linear relationship between the x and y values for the population.

9.11 Observed value: $t = 9.22$
The test is statistically significant at the 5% level, as P-value $\le (1/2)(.0001) < .05$.

9.13 (a) $2 \pm (3.143)\left(\dfrac{1.15}{\sqrt{168-(32)^2/8}}\right)$ or $1.43 < \beta_1 < 2.57$

(b) $-1.25 \pm (2.132)\left(\dfrac{.68}{\sqrt{232-(36)^2/6}}\right)$ or $-1.61 < \beta_1 < -.89$

(c) $-8.36459 \pm (2.33)(.6631)$ or $-9.910 < \beta_1 < -6.820$

9.15 $H_0: \rho = 0$; $H_a: \rho \neq 0$ $\alpha = .05$ $n = 8$

Observed value: $r = \dfrac{8(221)-(32)(62)}{\sqrt{8(134)-(32)^2}\,\sqrt{8(608)-(62)^2}} = -.976$

Critical values: $r = \pm.707$ Decision: Reject H_0
The data indicate a linear relationship between the x and y values for the population.

9.17 $H_0: \rho = 0$; $H_a: \rho \neq 0$ $\alpha = .05$ $n = 4$

Observed value: $r = \dfrac{4(58)-(12)(18)}{\sqrt{4(56)-(12)^2}\,\sqrt{4(86)-(18)^2}} = .400$

Critical values: $r = \pm.950$ Decision: Do not reject H_0
The data do not suggest a linear relationship between the x and y values for the population.

9.19 $H_0: \rho = 0$; $H_a: \rho \neq 0$ $\alpha = .05$ $n = 7$

Observed value: $r = \dfrac{7(20.428619)-(38.383)(3.721)}{\sqrt{7(211.279763)-(38.383)^2}\,\sqrt{7(1.983199)-(3.721)^2}} = .388$

Critical values: $r = \pm.754$ Decision: Do not reject H_0
The data do not indicate a linear relationship between the x and y values for the population.

9.21 $H_0: \rho = 0; H_a: \rho \neq 0 \qquad \alpha = .01 \qquad n = 9$

Observed value: $r = \dfrac{9(7439.37) - (41.56)(1416.1)}{\sqrt{9(289.4222) - (41.56)^2}\sqrt{9(232,498.97) - (1416.1)^2}} = .926$

Critical values: $r = \pm.798$ \qquad Decision: Reject H_0

The data indicate a linear relationship between the x and y values for the population.

9.23 $H_0: \rho = 0; H_a: \rho \neq 0 \qquad \alpha = .05 \qquad n = 11 \qquad$ Observed value: $r = .844416$

Critical values: $r = \pm.602$ \quad Decision: Reject H_0

The data indicate a linear relationship between the x and y values for the population.

9.25 (a) $\quad r > 0$ because $b_1 = .4686674 > 0 \qquad r = \sqrt{.10983} = .331$

(b) $\quad H_0: \rho = 0; H_a: \rho \neq 0 \quad \alpha = .05 \quad n = 40$

Observed value: $r = .331$ \qquad Critical values: $r = \pm.312$ \qquad Decision: Reject H_0

The data indicate a linear relationship between the x and y values for the population.

9.27 Linearity and Independence \qquad **9.29** No apparent violation

9.31 (a)

x	y	$\hat{y}$	$y - \hat{y}$	$(y - \hat{y})^2$	x^2	y^2	xy
4	2	3.98	−1.98	3.9204	16	4	8
4	2	3.98	−1.98	3.9204	16	4	8
4	5	3.98	1.02	1.0404	16	25	20
6	6	5.16	.84	.7056	36	36	36
8	7	6.34	.66	.4356	64	49	56
8	9	6.34	2.66	7.0756	64	81	72
15	9	10.47	−1.47	2.1609	225	81	135

Sums: $\quad$ 49 $\quad$ 40 $\qquad\qquad\qquad\qquad\quad$ 19.2589 $\quad$ 437 $\quad$ 280 $\quad$ 335

$n = 7 \qquad \bar{x} = 49/7 = 7 \qquad \bar{y} = 40/7 \qquad b_1 = \dfrac{7(335) - (49)(40)}{7(437) - (49)^2} = .585106$

$b_0 = (40/7) - (.585106)(7) = 1.62 \qquad \hat{y} = 1.62 + .59x$

(b) $\quad \hat{y}_0 = 1.62 + (.59)(5) = 4.57$

(c) $\quad 4.57 \pm (2.571)\sqrt{\dfrac{19.2589}{7-2}}\sqrt{1 + \dfrac{1}{7} + \dfrac{(5-7)^2}{437 - (49)^2/7}}$, or −.92 to 10.06

(d) $\quad 4.57 \pm (2.571)\sqrt{\dfrac{19.2589}{7-2}}\sqrt{\dfrac{1}{7} + \dfrac{(5-7)^2}{437 - (49)^2/7}}$, or 2.40 to 6.74

9.33 (a) $\quad H_0: \rho = 0; H_a: \rho \neq 0 \qquad \alpha = .05 \qquad n = 30 \qquad$ Observed value: $r = -.777$

Critical values: $r = \pm.361$ \qquad Decision: Reject H_0 \qquad The data suggest that there is a linear relationship between the x and y values for the population.

(b) $\quad \hat{y}_0 = 56.90 - (.31)(90) = 29 \qquad \bar{x} = 2444/30$

$29 \pm (2.048)\sqrt{\dfrac{315.11}{30-2}}\sqrt{1 + \dfrac{1}{30} + \dfrac{(90 - 2444/30)^2}{204,232 - (2444)^2/30}}$, or 21.97 to 36.03

(c) $\quad 29 \pm (2.048)\sqrt{\dfrac{315.11}{30-2}}\sqrt{\dfrac{1}{30} + \dfrac{(90 - 2444/30)^2}{204,232 - (2444)^2/30}}$, or 27.50 to 30.50

9.35   (a) $H_0: \beta_1 = 0$; $H_a: \beta_1 > 0$ $\alpha = .01$ $n = 6$

Observed value: $t = .9712699 \sqrt{\dfrac{6-2}{1-(.9712699)^2}} = 8.16$ df $= 6 - 2 = 4$

Critical value: $t = 3.747$ Decision: Reject H_0 P-value $< .005$ The data indicate a positive linear relationship between the x and y values for the population.

   (b)   In Exercise 3.7, the line of best fit was found to be $\hat{y} = -15.474 + 2.355x$.

$\hat{y}_0 = -15.474 + (2.355)(20) = 31.626$ per 100,000

9.37   (a) $H_0: \beta_1 = 0$; $H_a: \beta_1 > 0$ $\alpha = .05$ $n = 7$

$r = \dfrac{7(410.14) - (49.5)(54.6)}{\sqrt{7(386.79) - (49.5)^2}\sqrt{7(444.94) - (54.6)^2}} = .908278$

Observed value: $t = .908278 \sqrt{\dfrac{7-2}{1-(.908278)^2}} = 4.85$ df $= 7 - 2 = 5$

Critical value: $t = 2.015$ Decision: Reject H_0 P-value $< .005$ The data indicate a positive linear relationship between the x and y values for the population.

   (b) $\hat{y}_0 = 3.175 + (.654)(7) = 7.8$

9.39   (a) $H_0: \rho = 0$; $H_a: \rho \neq 0$ $\alpha = .05$ $n = 7$

Observed value: $r = .893$

Critical values: $r = \pm .754$

Decision: Reject H_0

The data suggest that there is a linear relationship between the x and y values for the population.

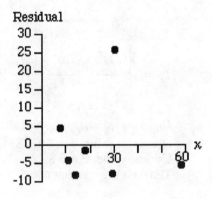

   (b)   From Exercise 3.49, $\hat{y} = -2.93 + 1.37x$. The residuals are given by $y - \hat{y}$. The first observation is $x = 59$ and $y = 72$. So the corresponding residual is $72 - [-2.93 + (1.37)(59)] = -5.900$. The plot of the residuals versus x indicate a large residual. This suggests that there may be a nonlinear relationship between x and y.

9.41   (a) P-value $= .168175 \geq .10$ Do not reject H_0 at the 10% significance level.

   (b)   Use $14 - 2 = 12$ degrees of freedom

$0.148406 \pm (1.782)(0.101185)$ or $-.032 < \beta_1 < .329$

The 90% confidence interval contains 0.

9.43   (a) P-value $= 0.314651 > .10$ Do not reject H_0 at the 10% significance level.

   (b)   Use $14 - 2 = 12$ degrees of freedom

$0.102196 \pm (1.782)(0.0973812)$ or $-.071 < \beta_1 < .276$

The 90% confidence interval contains 0.

9.45 $\hat{y}_0 = 66.29 - (.60)(70) = 24.29$ $n = 6$ df $= 6 - 2 = 4$

$\sum x = 460$ $\sum x^2 = 35,694$ $\bar{x} = 460/6$

   (a) $24.29 \pm (4.604)\sqrt{\dfrac{38.234}{6-2}}\sqrt{1 + \dfrac{1}{6} + \dfrac{(70 - 460/6)^2}{35,694 - (460)^2/6}}$, or 8.24 to 40.34

(b) $24.29 \pm (4.604)\sqrt{\dfrac{38.234}{6-2}}\sqrt{\dfrac{1}{6}+\dfrac{(70-460/6)^2}{35,694-(460)^2/6}}$, or 16.88 to 31.70

Chapter 10

10.1

	$F(.01)$		$F(.025)$		$F(.05)$		$F(.01)$		$F(.025)$		$F(.05)$
(ai)	6.63	(aii)	4.82	(aiii)	3.69	(bi)	10.29	(bii)	6.76	(biii)	4.82
(ci)	4.41	(cii)	3.42	(ciii)	2.77	(di)	12.25	(dii)	8.07	(diii)	5.59

10.3

	Observed F value		df	$F(\alpha)$	Decision
(a)	$5(50)/(100/3)$	$= 7.5$	(2, 12)	3.89	Reject H_0
(b)	$5(20)/(100/3)$	$= 3$	(2, 12)	3.89	Do not reject H_0
(c)	$6(25)/(50/2)$	$= 6$	(1, 10)	10.04	Do not reject H_0
(d)	$6(25)/(25/2)$	$= 12$	(1, 10)	10.04	Reject H_0
(e)	$4(40)/(200/5)$	$= 4$	(4, 15)	3.06	Reject H_0
(f)	$4(40)/(400/5)$	$= 2$	(4, 15)	3.06	Do not reject H_0

10.5

	A	B	C				
$\bar{x}$:	22	25	27	$m = 5$	$k = 3$	s^2:	6.5 5.5 8

$s_{\bar{x}}^2 = \dfrac{3\left[(22)^2+(25)^2+(27)^2\right]-(22+25+27)^2}{3(2)} = 6.333$

$s_A^2 + s_B^2 + s_C^2 = 6.5 + 5.5 + 8 = 20$

$H_0: \mu_A = \mu_B = \mu_C$; H_a: Not all the means are equal. $\alpha = .05$

Observed value: $F = 5(6.333)/(20/3) = 4.75$ df $= (2, 12)$

Critical value: $F = 3.89$ Decision: Reject H_0 $.025 < P\text{-value} < .05$

The data suggest that not all the means are the same.

10.7 $s_{\bar{x}}^2 = \dfrac{2[(180.3)^2+(190.8)^2]-(180.3+190.8)^2}{2(1)} = 55.125$

$s_M^2 + s_F^2 = (7.2)^2 + (10.7)^2 = 166.33$ $H_0: \mu_M = \mu_F$; H_a: $\mu_M \neq \mu_F$ $\alpha = .01$

Observed value: $F = 10(55.125)/(166.33/2) = 6.63$ df $= (1, 18)$

Critical value: $F = 8.29$ Decision: Do not reject H_0 $.01 < P\text{-value} < .025$

The data do not suggest that the means are different.

10.9 $s_{\bar{x}}^2 = \dfrac{5\left[(24)^2+(25)^2+(25)^2+(30)^2+(22)^2\right]-(24+25+25+30+22)^2}{5(4)} = 8.7$

$s_A^2 + s_B^2 + s_C^2 + s_D^2 + s_E^2 = (5.02)^2 + (4.69)^2 + (5.66)^2 + (4.38)^2 + (4.65)^2 = 120.039$

$H_0: \mu_A = \mu_B = \mu_C = \mu_D = \mu_E$; H_a: Not all the means are equal. $\alpha = .05$

(ai) $k = 5$ $n = 30$ df $= (k-1, n-k) = (4, 25)$ (aiii) $F = 2.76$

(aiii) $F = \dfrac{6(8.7)}{(120.039)/5} = 2.17$ (aiv) Decision: Do not reject H_0 $P\text{-value} > .05$

(bi) Same   (bii) Larger   (biii) Larger

(biv) The sample variances within each group remain the same. The only change in sample means is that $\bar{x}_E = 18$.

$$s_{\bar{x}}^2 = \frac{5\left[(24)^2 + (25)^2 + (25)^2 + (30)^2 + (18)^2\right] - (24 + 25 + 25 + 30 + 18)^2}{5(4)} = 18.3$$

$$F = \frac{6(18.3)}{(120.039)/5} = 4.57 \qquad \text{Decision: Reject } H_0 \qquad P\text{-value} < .01$$

10.11 (a) $H_0: \mu_1 = \mu_2 = \mu_3$; H_a: Not all the means are equal.

(b) Observed value: $F = 43.79$ The test is statistically significant, as P-value $= .000$.

10.13 $\Sigma x_A = 81 \qquad \bar{x}_A = 27 \qquad \Sigma(x_A - \bar{x}_A)^2 = 18 \qquad n_A = 3$

$\Sigma x_B = 175 \quad \bar{x}_B = 35 \qquad \Sigma(x_B - \bar{x}_B)^2 = 58 \qquad n_B = 5$

$SSW = 18 + 58 = 76 \qquad MSW = 76/(8 - 2) = 12.667$

$SSB = 3(27 - 32)^2 + 5(35 - 32)^2 = 120$

$\bar{\bar{x}} = (81 + 175)/8 = 32 \qquad MSB = 120/(2 - 1) = 120$

Source	SS	df	MS	F statistic
Between samples	120	1	120	9.47
Within samples	76	6	12.667	
Total	196	7		

$H_0: \mu_A = \mu_B$; $H_a: \mu_A \neq \mu_B$ $\alpha = .05$ Observed value: $F = 120/12.667 = 9.47$
df $= (1, 6)$ Critical value: $F = 5.99$ Decision: Reject H_0 $.01 < P\text{-value} < .025$
The data suggest that the means are different.

10.15 $\Sigma x_A = 36 \qquad \bar{x}_A = 12 \qquad \Sigma(x_A - \bar{x}_A)^2 = 24 \qquad n_A = 3$

$\Sigma x_B = 18 \qquad \bar{x}_B = 6 \qquad \Sigma(x_B - \bar{x}_B)^2 = 18 \qquad n_B = 3$

$\Sigma x_C = 28 \qquad \bar{x}_C = 7 \qquad \Sigma(x_C - \bar{x}_C)^2 = 20 \qquad n_C = 4$

$SSW = 24 + 18 + 20 = 62 \quad MSW = 62/(10 - 3) = 8.857 \qquad \bar{\bar{x}} = (36 + 18 + 28)/10 = 8.2$

$SSB = 3(12 - 8.2)^2 + 3(6 - 8.2)^2 + 4(7 - 8.2)^2 = 63.6 \quad MSB = 63.6/(3 - 1) = 31.8$

Source	SS	df	MS	F statistic
Between samples	63.6	2	31.8	3.59
Within samples	62	7	8.857	
Total	125.6	9		

$H_0: \mu_A = \mu_B = \mu_C$; H_a: Not all the means are equal. $\alpha = .05$ Critical value: $F = 4.74$
Observed value: $F = 31.8/8.857 = 3.59$ df $= (2, 7)$
Decision: Do not reject H_0 $P\text{-value} > .05$
The data do not suggest that there is a difference in means.

10.17 (a)

Source	SS	df	MS	F statistic
Between samples	250	5	50	8.33
Within samples	150	25	6	
Total	400	30		

(b) H_0: $\mu_A = \mu_B = \mu_C = \mu_D = \mu_E = \mu_F$; H_a: Not all the means are equal. $\alpha = .05$
Observed value: $F = 50/6 = 8.33$ df = (5, 25)   Critical value: $F = 2.60$
Decision: Reject H_0 P-value $< .01$
The data suggest that not all the means are the same.

10.19 (a)

Source	SS	df	MS	F statistic
Between samples	128	4	32	2
Within samples	160	10	16	
Total	288	14		

(b) H_0: $\mu_A = \mu_B = \mu_C = \mu_D = \mu_E$; H_a: Not all the means are equal. $\alpha = .05$
Observed value: $F = 32/16 = 2$ df = (4, 10)   Critical value: $F = 3.48$
Decision: Do not reject H_0 P-value $> .05$
The data do not suggest that there is a difference in means.

10.21 H_0: $\mu_A = \mu_B = \mu_C$; H_a: Not all the means are equal. $\alpha = .05$
$SSW = 4(12.806)^2 + 5(12.215)^2 + 7(11.637)^2 = 2349.944$
$MSW = 2349.944/(19 - 3) = 146.872$ $\bar{\bar{x}} = [5(66) + 6(68) + 8(83)]/19 = 73.789$
$SSB = 5(66 - 73.789)^2 + 6(68 - 73.789)^2 + 8(83 - 73.789)^2 = 1183.158$
$MSB = 1183.158/(3 - 1) = 591.579$ Observed value: $F = 591.579/146.872 = 4.03$
df = (2, 16)   Critical value: $F = 3.63$ Decision: Reject H_0 $.025 < P$-value $< .05$
The data suggest that not all the means are the same.

10.23 H_0: $\mu_P = \mu_E = \mu_W = \mu_S$; H_a: Not all the means are equal. $\alpha = .05$
$SSW = 3(.768)^2 + 4(.554)^2 + 5(.652)^2 + 8(.970)^2 = 12.650$
$MSW = 12.650/(24 - 4) = .633$
$\bar{\bar{x}} = [4(6.775) + 5(7.22) + 6(6.017) + 9(5.644)]/24 = 6.254$
$SSB = 4(6.775 - 6.254)^2 + 5(7.22 - 6.254)^2 + 6(6.017 - 6.254)^2 + 9(5.644 - 6.254)^2$
$\quad = 9.437$ $MSB = 9.437/(4 - 1) = 3.146$ Observed value: $F = 3.146/.633 = 4.97$
df = (3, 20)   Critical value: $F = 3.10$ Decision: Reject H_0 P-value $< .01$
The data suggest that not all the means are the same.

10.25 H_0: $\mu_{CR} = \mu_M = \mu_{CO}$; H_a: Not all the means are equal. $\alpha = .01$
$SSW = 27(4.5)^2 + 33(4.6)^2 + 33(5.2)^2 = 2137.350$
$MSW = 2137.350/(96 - 3) = 22.982$
$\bar{\bar{x}} = [28(26.5) + 34(25.5) + 34(31.2)]/96 = 27.810$
$SSB = 28(26.5 - 27.81)^2 + 34(25.5 - 27.81)^2 + 34(31.2 - 27.81)^2 = 620.210$
$MSB = 620.21/(3 - 1) = 310.105$ Observed value: $F = 310.105/22.982 = 13.49$
df = (2, 93)   Critical value: $F = 4.98$
(Note that we used df = (2,60) in getting a critical value.)
Decision: Reject H_0 P-value $< .01$ The data suggest that not all the means are the same.

10.27 (a) $T_A = 81$ $n_A = 3$ $T_B = 175$ $n_B = 5$ $SSB = \dfrac{(81)^2}{3} + \dfrac{(175)^2}{5} - \dfrac{(81+175)^2}{8} = 120$

$SSW = (30)^2 + (27)^2 + (24)^2 + (35)^2 + (37)^2 + (33)^2 + (30)^2 + (40)^2$
$\quad - \left(\dfrac{(81)^2}{3} + \dfrac{(175)^2}{5} \right) = 76$

(b) $T_A = 128$ $n_A = 4$ $T_B = 100$ $n_B = 4$

$$SSB = \frac{(128)^2}{4} + \frac{(100)^2}{4} - \frac{(128+100)^2}{8} = 98$$

$$SSW = (29)^2 + (38)^2 + (25)^2 + (36)^2 + (26)^2 + (31)^2 + (25)^2 + (18)^2$$
$$- \left(\frac{(128)^2}{4} + \frac{(100)^2}{4} \right) = 196$$

(c) $T_A = 36$ $n_A = 3$ $T_B = 18$ $n_B = 3$ $T_C = 28$ $n_C = 4$

$$SSB = \frac{(36)^2}{3} + \frac{(18)^2}{3} + \frac{(28)^2}{4} - \frac{(36+18+28)^2}{10} = 63.6$$

$$SSW = (10)^2 + (16)^2 + (10)^2 + (9)^2 + (3)^2 + (6)^2 + (8)^2 + (10)^2 + (4)^2 + (6)^2$$
$$- \left(\frac{(36)^2}{3} + \frac{(18)^2}{3} + \frac{(28)^2}{4} \right) = 62$$

(d) $T_A = 18$ $n_A = 3$ $T_B = 12$ $n_B = 2$ $T_C = 32$ $n_C = 4$ $T_D = 80$ $n_D = 5$

$$SSB = \frac{(18)^2}{3} + \frac{(12)^2}{2} + \frac{(32)^2}{4} + \frac{(80)^2}{5} - \frac{(18+12+32+80)^2}{14} = 275.714$$

$$SSW = (12)^2 + (2)^2 + (4)^2 + (10)^2 + (2)^2 + (6)^2 + (6)^2 + (14)^2 + (6)^2 + (22)^2$$
$$+ (16)^2 + (16)^2 + (13)^2 + (13)^2 \ - \left(\frac{(18)^2}{3} + \frac{(12)^2}{2} + \frac{(32)^2}{4} + \frac{(80)^2}{5} \right) = 190$$

10.29 $\sum x_A = 20$ $\bar{x}_A = 10$ $\sum(x_A - \bar{x}_A)^2 = 50$ $n_A = 2$

$\sum x_B = 60$ $\bar{x}_B = 20$ $\sum(x_B - \bar{x}_B)^2 = 50$ $n_B = 3$

$\sum x_C = 125$ $\bar{x}_C = 25$ $\sum(x_C - \bar{x}_C)^2 = 100$ $n_C = 5$

$\bar{\bar{x}} = (20 + 60 + 125)/10 = 20.5$ $SSW = 50 + 50 + 100 = 200$

$MSW = 200/(10 - 3) = 28.571$

$SSB = 2(10 - 20.5)^2 + 3(20 - 20.5)^2 + 5(25 - 20.5)^2 = 322.5$

$MSB = 322.5/(3 - 1) = 161.25$

$F = 161.25/28.571 = 5.64$

Source	SS	df	MS	F statistic
Between samples	322.5	2	161.25	5.64
Within samples	200	7	28.571	
Total	522.5	9		

$H_0: \mu_A = \mu_B = \mu_C$; H_a: Not all the means are equal. $\alpha = .05$ Observed value: $F = 5.64$
df $= (2, 7)$ Critical value: $F = 4.74$ Decision: Reject H_0 $.025 < P$-value $< .05$
The data suggest that not all the means are the same.

10.31 (a)

Source	SS	df	MS	F statistic
Between samples	132	4	33	3.3
Within samples	400	40	10	
Total	532	44		

(b) $H_0: \mu_A = \mu_B = \mu_C = \mu_D = \mu_E$; H_a: Not all the means are equal. $\alpha = .05$
Observed value: $F = 33/10 = 3.3$ df $= (4, 40)$ Critical value: $F = 2.61$
Decision: Reject H_0 $.01 < P$-value $< .025$
The data suggest that not all the means are the same.

10.33 $\Sigma x_A = 240$ $\bar{x}_A = 60$ $\Sigma(x_A - \bar{x}_A)^2 = 288$ $n_A = 4$

$\Sigma x_B = 162$ $\bar{x}_B = 54$ $\Sigma(x_B - \bar{x}_B)^2 = 86$ $n_B = 3$

$\Sigma x_C = 160$ $\bar{x}_C = 40$ $\Sigma(x_C - \bar{x}_C)^2 = 200$ $n_C = 4$

$\bar{\bar{x}} = (240 + 162 + 160)/11 = 51.091$ $SSW = 288 + 86 + 200 = 574$
$MSW = 574/(11 - 3) = 71.75$
$SSB = 4(60 - 51.091)^2 + 3(54 - 51.091)^2 + 4(40 - 51.091)^2 = 834.909$
$MSB = 834.909/(3 - 1) = 417.455$
$H_0: \mu_A = \mu_B = \mu_C$; H_a: Not all the means are equal. $\alpha = .05$
Observed value: $F = 417.455/71.75 = 5.82$ df $= (2, 8)$ Critical value: $F = 4.46$
Decision: Reject H_0 $.025 < P$-value $< .05$
The data suggest that not all the means are the same.

10.35 $\Sigma x_A = 14.4$ $\bar{x}_A = 2.4$ $\Sigma(x_A - \bar{x}_A)^2 = .44$ $n_A = 6$

$\Sigma x_B = 16.8$ $\bar{x}_B = 2.8$ $\Sigma(x_B - \bar{x}_B)^2 = .44$ $n_B = 6$

$\Sigma x_C = 15$ $\bar{x}_C = 3$ $\Sigma(x_C - \bar{x}_C)^2 = .34$ $n_C = 5$

$\bar{\bar{x}} = (14.4 + 16.8 + 15)/17 = 2.718$ $SSW = .44 + .44 + .34 = 1.22$
$MSW = 1.22/(17 - 3) = .087$
$SSB = 6(2.4 - 2.718)^2 + 6(2.8 - 2.718)^2 + 5(3 - 2.718)^2 = 1.045$
$MSB = 1.045/(3 - 1) = .523$ $H_0: \mu_A = \mu_B = \mu_C$; H_a: Not all the means are equal. $\alpha = .05$
Observed value: $F = .523/.087 = 6.01$ df $= (2, 14)$ Critical value: $F = 3.74$
Decision: Reject H_0 $.01 < P$-value $< .025$
The data suggest that not all the means are the same.

10.37 $\Sigma x_A = 180$ $\bar{x}_A = 36$ $\Sigma(x_A - \bar{x}_A)^2 = 320$ $n_A = 5$

$\Sigma x_B = 190$ $\bar{x}_B = 38$ $\Sigma(x_B - \bar{x}_B)^2 = 200$ $n_B = 5$

$\Sigma x_C = 250$ $\bar{x}_C = 50$ $\Sigma(x_C - \bar{x}_C)^2 = 280$ $n_C = 5$

$\bar{\bar{x}} = (180 + 190 + 250)/15 = 41.333$ $SSW = 320 + 200 + 280 = 800$
$MSW = 800/(15 - 3) = 66.667$
$SSB = 5(36 - 41.333)^2 + 5(38 - 41.333)^2 + 5(50 - 41.333)^2 = 573.333$
$MSB = 573.333/(3 - 1) = 286.667$ $H_0: \mu_A = \mu_B = \mu_C$; H_a: Not all the means are equal.
$\alpha = .05$ Observed value: $F = 286.667/66.667 = 4.30$ df $= (2, 12)$
Critical value: $F = 3.89$ Decision: Reject H_0 $.025 < P$-value $< .05$
The data suggest that not all the means are the same.

10.39 H_0: $\mu_E = \mu_G = \mu_N$; H_a: Not all the means are equal. $\alpha = .05$

$SSW = 19(24.5)^2 + 7(29.5)^2 + 71(35.6)^2 = 107{,}479.06$
$MSW = 107{,}479.06/(100 - 3) = 1108.032$
$\bar{\bar{x}} = [20(108) + 8(121) + 72(124)]/100 = 120.56$
$SSB = 20(108 - 120.56)^2 + 8(121 - 120.56)^2 + 72(124 - 120.56)^2 = 4008.64$
$MSB = 4008.64/(3 - 1) = 2004.32$ Observed value: $F = 2004.32/1108.032 = 1.81$
df = (2, 97)   Critical value: $F = 3.15$
(Note that we used df = (2,60) in getting a critical value.)
Decision: Do not reject H_0 P-value > .05
The data do not suggest that there is a difference in means.

10.41
(a)

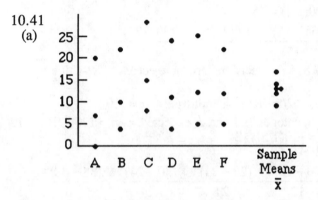

(b) $\sum x_A = 27$ $\bar{x}_A = 9$

$\sum (x_A - \bar{x}_A)^2 = 206$ $n_A = 3$
$\sum x_B = 36$ $\bar{x}_B = 12$
$\sum (x_B - \bar{x}_B)^2 = 168$ $n_B = 3$
$\sum x_C = 51$ $\bar{x}_C = 17$
$\sum (x_C - \bar{x}_C)^2 = 206$ $n_C = 3$
$\sum x_D = 39$ $\bar{x}_D = 13$
$\sum (x_D - \bar{x}_D)^2 = 206$ $n_D = 3$
$\sum x_E = 42$ $\bar{x}_E = 14$
$\sum (x_E - \bar{x}_E)^2 = 206$ $n_E = 3$
$\sum x_F = 39$ $\bar{x}_F = 13$
$\sum (x_F - \bar{x}_F)^2 = 146$ $n_F = 3$

$\bar{\bar{x}} = (27 + 36 + 51 + 39 + 42 + 39)/18 = 13$
$SSW = 206 + 168 + 206 + 206 + 206 + 146 = 1138$
$MSW = 1138/(18 - 6) = 94.833$
$SSB = 3(9 - 13)^2 + 3(12 - 13)^2 + 3(17 - 13)^2 + 3(13 - 13)^2 + 3(14 - 13)^2$
$\qquad + 3(13 - 13)^2 = 102$ $MSB = 102/(6 - 1) = 20.4$
H_0: $\mu_A = \mu_B = \mu_C = \mu_D = \mu_E = \mu_F$; H_a: Not all the means are equal. $\alpha = .05$
Observed value: $F = 20.4/94.833 = .22$ df = (5, 12)   Critical value: $F = 3.11$
Decision: Do not reject H_0 P-value > .05   The data do not suggest that there is a
difference in means.

10.43   (a)   The ANOVA table is identical with the table in Exercise 10.42.
        (b)   When 2 is added to each data value, the variance of the means and the variance within
              each sample remains the same.

10.45 H_0: $\mu_A - \mu_B = 0$; H_a: $\mu_A - \mu_B \neq 0$ $\alpha = .05$
        Observed value: $F = 1.72$ df = (1, 30)   Critical value: $F = 4.17$
        Decision: Do not reject H_0 P-value = .20
        The data do not suggest that there is a difference in mean final exam scores between the two
        sections.

Chapter 11

11.1   (ai)   21.666   (aii)   19.023   (aiii) 16.919   (aiv) 3.325   (av)   2.088
        (bi)   30.578   (bii)   27.488   (biii) 24.996   (biv) 7.261   (bv)   5.229
        (ci)   44.314   (cii)   40.646   (ciii) 37.652   (civ) 14.611   (cv) 11.524

11.3   (a)   11   (b)   .95   (c)   3.940   (d)   25   (e)   .10   (f)   27.587
       (g)   12.443   (h)   3.053

11.5 H_0: $p_1 = p_2 = p_3 = p_4 = 1/4$; H_a: H_0 is not true. $\alpha = .05$ $n = 68 + 65 + 77 + 90 = 300$
In each category, expected $= np = 300(1/4) = 75$

Observed value: $\chi^2 = \dfrac{(68-75)^2}{75} + \dfrac{(65-75)^2}{75} + \dfrac{(77-75)^2}{75} + \dfrac{(90-75)^2}{75} = 5.04$

df $= 4 - 1 = 3$ Critical value: $\chi^2 = 7.815$ Decision: Do not reject H_0 P-value $> .10$

11.7 H_0: $p_1 = p_2 = 1/10$, $p_3 = p_4 = 1/5$, $p_5 = 2/5$; H_a: H_0 is not true. $\alpha = .10$
$n = 12 + 20 + 20 + 30 + 38 = 120$ In categories 1 and 2, expected $= np = 120(1/10) = 12$
In categories 3 and 4, expected $= np = 120(1/5) = 24$
In category 5, expected $= np = 120(2/5) = 48$

Observed value: $\chi^2 = \dfrac{(12-12)^2}{12} + \dfrac{(20-12)^2}{12} + \dfrac{(20-24)^2}{24} + \dfrac{(30-24)^2}{24} + \dfrac{(38-48)^2}{48}$

$= 9.58$ df $= 5 - 1 = 4$ Critical value: $\chi^2 = 7.779$
Decision: Reject H_0 $.025 < P$-value $< .05$

11.9   Let A, B, C, and D represent white women, minority women, white men, and
minority men, respectively.
H_0: $p_A = .30$, $p_B = .05$, $p_C = .50$, $p_D = .15$; H_a: H_0 is not true. $\alpha = .05$
$n = 40 + 15 + 80 + 15 = 150$ In category A, expected $= np = 150(.30) = 45$
In category B, expected $= np = 150(.05) = 7.5$
In category C, expected $= np = 150(.50) = 75$
In category D, expected $= np = 150(.15) = 22.5$

Observed value: $\chi^2 = \dfrac{(40-45)^2}{45} + \dfrac{(15-7.5)^2}{7.5} + \dfrac{(80-75)^2}{75} + \dfrac{(15-22.5)^2}{22.5} = 10.89$

df $= 4 - 1 = 3$ Critical value: $\chi^2 = 7.815$ Decision: Reject H_0 $.01 < P$-value $< .025$
The data suggest that the official's claim is not true.

11.11 Let A, B, C, and D represent less than $25,000, between $25,000 and $30,000, between
$30,000 and $35,000, and more than $35,000 respectively.
H_0: $p_A = 1/10$ $p_B = p_C = p_D = 3/10$; H_a: H_0 is not true. $\alpha = .10$
$n = 19 + 56 + 51 + 40 = 166$ In category A, expected $= np = 166(1/10) = 16.6$
In the other categories, expected $= np = 166(3/10) = 49.8$

Observed value: $\chi^2 = \dfrac{(19-16.6)^2}{16.6} + \dfrac{(56-49.8)^2}{49.8} + \dfrac{(51-49.8)^2}{49.8} + \dfrac{(40-49.8)^2}{49.8} = 3.08$

df $= 4 - 1 = 3$ Critical value: $\chi^2 = 6.251$ Decision: Do not reject H_0 P-value $> .10$
The data do not dispute the owner's assertion.

11.13 H_0: $p = 1/10$ for each category; H_a: H_0 is not true. $\alpha = .05$
$n = 6 + 6 + 13 + 9 + 13 + 11 + 8 + 12 + 10 + 12 = 100$
In each category, expected $= np = 100(1/10) = 10$
Observed value:

$$\chi^2 = \frac{(6-10)^2}{10} + \frac{(6-10)^2}{10} + \frac{(13-10)^2}{10} + \frac{(9-10)^2}{10} + \frac{(13-10)^2}{10} + \frac{(11-10)^2}{10}$$

$$+ \frac{(8-10)^2}{10} + \frac{(12-10)^2}{10} + \frac{(10-10)^2}{10} + \frac{(12-10)^2}{10} = 6.40$$

df $= 10 - 1 = 9$ Critical value: $\chi^2 = 16.919$ Decision: Do not reject H_0 P-value $> .10$
The data do not dispute the claim.

11.15 H_0: Characteristics of ratings and age are independent.
H_a: Characteristics of ratings and age are related. $\alpha = .05$
Expected values: $(26)(16)/50 = 8.32$ $(26)(18)/50 = 9.36$ $(26)(16)/50 = 8.32$
                    $(24)(16)/50 = 7.68$ $(24)(18)/50 = 8.64$ $(24)(16)/50 = 7.68$

Observed value:

$$\chi^2 = \frac{(6-8.32)^2}{8.32} + \frac{(12-9.36)^2}{9.36} + \frac{(8-8.32)^2}{8.32} + \frac{(10-7.68)^2}{7.68} + \frac{(6-8.64)^2}{8.64}$$

$$+ \frac{(8-7.68)^2}{7.68} = 2.92 \quad \text{df} = (2-1)(3-1) = 2 \quad \text{Critical value: } \chi^2 = 5.991$$

Decision: Do not reject H_0 P-value $> .10$
The data do not suggest that the characteristics of rating and age are related.

11.17 H_0: Characteristics of gender and CHD are independent.
H_a: Characteristics of gender and CHD are related. $\alpha = .05$
Expected values: $(11)(15)/28 = 5.89$ $(11)(13)/28 = 5.11$
                    $(17)(15)/28 = 9.11$ $(17)(13)/28 = 7.89$

Observed value: $\chi^2 = \dfrac{(5-5.89)^2}{5.89} + \dfrac{(6-5.11)^2}{5.11} + \dfrac{(10-9.11)^2}{9.11} + \dfrac{(7-7.89)^2}{7.89} = .48$

df $= (2-1)(2-1) = 1$ Critical value: $\chi^2 = 3.841$
Decision: Do not reject H_0 P-value $> .10$
The data do not suggest that the characteristics of gender and CHD are related.

11.19 H_0: Characteristics of sex and salaries are independent.
H_a: Characteristics of sex and salaries are related. $\alpha = .10$
Expected values:
 $(75)(19)/166 = 8.58$ $(75)(56)/166 = 25.3$ $(75)(51)/166 = 23.04$ $(75)(40)/166 = 18.07$
 $(91)(19)/166 = 10.42$ $(91)(56)/166 = 30.7$ $(91)(51)/166 = 27.96$ $(91)(40)/166 = 21.93$
Observed value:

$$\chi^2 = \frac{(12-8.58)^2}{8.58} + \frac{(30-25.3)^2}{25.3} + \frac{(20-23.04)^2}{23.04} + \frac{(13-18.07)^2}{18.07}$$

$$+ \frac{(7-10.42)^2}{10.42} + \frac{(26-30.7)^2}{30.7} + \frac{(31-27.96)^2}{27.96} + \frac{(27-21.93)^2}{21.93} = 7.40$$

df $= (2-1)(4-1) = 3$ Critical value: $\chi^2 = 6.251$
Decision: Reject H_0 $.05 < P$-value $< .10$
The data suggest that the characteristics of sex and salary are related.

11.21 (ai) Expected values: $(60)(30)/180 = 10$ $(60)(60)/180 = 20$ $(60)(90)/180 = 30$
$(120)(30)/180 = 20$ $(120)(60)/180 = 40$ $(120)(90)/180 = 60$

Notice that the observed equals the expected in each of the six cells. $\chi^2 = 0$
(aii) Expected values: $(180)(160)/240 = 120$ $(180)(80)/240 = 60$
$(60)(160)/240 = 40$ $(60)(80)/240 = 20$

Notice that the observed equals the expected in each of the four cells. $\chi^2 = 0$
(aiii) Expected values: $(50)(260)/650 = 20$ $(50)(390)/650 = 30$
$(100)(260)/650 = 40$ $(100)(390)/650 = 60$
$(500)(260)/650 = 200$ $(500)(390)/650 = 300$

Notice that the observed equals the expected in each of the six cells. $\chi^2 = 0$
(ci)   (cii)

	B_1	B_2	B_3		B_1	B_2	B_3
A_1	20	10	30	A_1	20	10	30
A_2	60	30	90	A_2	60	30	90
A_3	180	90	270	A_3	30	15	45

(d)   Rows (and columns) are proportional

11.23 H_0: The two dice are homogeneous with respect to the number of dots showing.
H_a: H_0 is not true $\alpha = .05$
Expected values:
$(35)(11)/70 = 5.5$ $(35)(10)/70 = 5$ $(35)(15)/70 = 7.5$ $(35)(11)/70 = 5.5$
$(35)(10)/70 = 5$ $(35)(13)/70 = 6.5$ $(35)(11)/70 = 5.5$ $(35)(10)/70 = 5$
$(35)(15)/70 = 7.5$ $(35)(11)/70 = 5.5$ $(35)(10)/70 = 5$ $(35)(13)/70 = 6.5$
Observed value:

$$\chi^2 = \frac{(4-5.5)^2}{5.5} + \frac{(6-5)^2}{5} + \frac{(7-7.5)^2}{7.5} + \frac{(5-5.5)^2}{5.5} + \frac{(5-5)^2}{5} + \frac{(8-6.5)^2}{6.5}$$

$$+ \frac{(7-5.5)^2}{5.5} + \frac{(4-5)^2}{5} + \frac{(8-7.5)^2}{7.5} + \frac{(6-5.5)^2}{5.5} + \frac{(5-5)^2}{5} + \frac{(5-6.5)^2}{6.5} = 2.07$$

df $= (2-1)(6-1) = 5$ Critical value: $\chi^2 = 11.071$
Decision: Do not reject H_0 P-value $> .10$
There is insufficient evidence to conclude that the dice are not homogeneous.

11.25 H_0: The appropriate populations of males and females are homogeneous with
respect to belief of discrimination in salaries; H_a: H_0 is not true. $\alpha = .01$
Expected values: $(125)(120)/230 = 65.22$ $(125)(110)/230 = 59.78$
$(105)(120)/230 = 54.78$ $(105)(110)/230 = 50.22$

Observed value:

$$\chi^2 = \frac{(55-65.22)^2}{65.22} + \frac{(70-59.78)^2}{59.78} + \frac{(65-54.78)^2}{54.78} + \frac{(40-50.22)^2}{50.22} = 7.34$$

df $= (2-1)(2-1) = 1$ Critical value: $\chi^2 = 6.635$
Decision: Reject H_0 $.005 < P$-value $< .01$
The data suggest that the appropriate populations are not homogeneous.

11.29 $H_0: p_A = .10 \quad p_B = .20 \quad p_C = .30 \quad p_D = .25 \quad p_F = .15; H_a: H_0$ is not true. $\alpha = .05$
$n = 12 + 20 + 26 + 14 + 8 = 80$
In category A, expected $= np = 80(.10) = 8$ In category B, expected $= np = 80(.20) = 16$
In category C, expected $= np = 80(.30) = 24$ In category D, expected $= np = 80(.25) = 20$
In category F, expected $= np = 80(.15) = 12$

Observed value: $\chi^2 = \dfrac{(12-8)^2}{8} + \dfrac{(20-16)^2}{16} + \dfrac{(26-24)^2}{24} + \dfrac{(14-20)^2}{20} + \dfrac{(8-12)^2}{12} = 6.30$

df $= 5 - 1 = 4$ Critical value: $\chi^2 = 9.488$ Decision: Do not reject H_0 P-value $> .10$
The data do not suggest the new teacher's grading policy is different from that of the department.

11.31 Let A, B, C, and D represent the categories drive alone, carpool, public transportation, and other means respectively.
$H_0: p_A = .70 \quad p_B = .20 \quad p_C = .08 \quad p_D = .02; H_a: H_0$ is not true. $\alpha = .05$
$n = 320 + 130 + 35 + 15 = 500$
In category A, expected $= np = 500(.70) = 350$
In category B, expected $= np = 500(.20) = 100$
In category C, expected $= np = 500(.08) = 40$
In category D, expected $= np = 500(.02) = 10$

Observed value: $\chi^2 = \dfrac{(320-350)^2}{350} + \dfrac{(130-100)^2}{100} + \dfrac{(35-40)^2}{40} + \dfrac{(15-10)^2}{10} = 14.70$

df $= 4 - 1 = 3$ Critical value: $\chi^2 = 7.815$ Decision: Reject H_0 P-value $< .005$
The data suggest that the modes of transportation to work have changed.

11.33 H_0: Characteristics of gender and CHD are independent.
H_a: Characteristics of gender and CHD are related. $\qquad \alpha = .10$
Expected values: $(86)(93)/155 = 51.6$ $(86)(62)/155 = 34.4$
$\qquad\qquad\qquad (69)(93)/155 = 41.4$ $(69)(62)/155 = 27.6$
Observed value:

$$\chi^2 = \frac{(54-51.6)^2}{51.6} + \frac{(32-34.4)^2}{34.4} + \frac{(39-41.4)^2}{41.4} + \frac{(30-27.6)^2}{27.6} = .63$$

df $= (2-1)(2-1) = 1$ Critical value: $\chi^2 = 2.706$
Decision: Do not reject H_0 P-value $> .10$
The data do not suggest that the characteristics of gender and CHD are related.

11.35 H_0: The appropriate male and female populations are homogeneous with respect to age.
$H_a: H_0$ is not true. $\qquad \alpha = .05$
Expected values: $(75)(85)/175 = 36.43$ $(75)(60)/175 = 25.71$ $(75)(30)/175 = 12.86$
$\qquad\qquad\qquad (100)(85)/175 = 48.57$ $(100)(60)/175 = 34.29$ $(100)(30)/175 = 17.14$
Observed value:

$$\chi^2 = \frac{(45-36.43)^2}{36.43} + \frac{(20-25.71)^2}{25.71} + \frac{(10-12.86)^2}{12.86} + \frac{(40-48.57)^2}{48.57}$$

$$+ \frac{(40-34.29)^2}{34.29} + \frac{(20-17.14)^2}{17.14} = 6.86$$

df $= (2-1)(3-1) = 2$ Critical value: $\chi^2 = 5.991$
Decision: Reject H_0 $.025 < P$-value $< .05$ The data suggest that the male and female populations are not homogeneous with respect to age.

11.37 (a) H_0: The attitude concerning racial quotas in hiring are homogeneous with respect to appropriate populations of Democrats, Republicans, and Independents; H_a: H_0 is not true.

(b) Same

Chapter 12

12.1 (a) {0, 1, 2, 10, 11, 12}   (b) .055 + .008 = .063
(c) {0, 1, 2}   (d) .004 + .031 + .031 + .004 = .07
(e) {11, 12, 13, 14, 15}   (f) .002 + .010 + .035 = .047
(g) {0, 1, 2, 3, 4, 5}   (h) .027 + .005 = .032
(i) .062 + .062 = .124   (j) .031
(k) {0, 1, 2, 3, 4, 12, 13, 14, 15, 16}   (l) {14, 15, 16, 17, 18, 19}

12.3 H_0: Md = 2.8; H_a: Md > 2.8 α = .05
Test statistic and observed value:
Let x be the number of data values larger than 2.8. $x = 8$
Critical region: {8, 9, 10} $n = 10$ α = .044 + .010 + .001 = .055
Decision: Reject H_0 The data support the manager's claim.

12.5 H_0: Md = 0; H_a: Md > 0 α = .05
Test statistic and observed value:
Let x be the number of differences (weight before − weight after) larger than 0. $x = 6$
Critical region: {7, 8} $n = 8$ α = .031 + .004 = .035
Decision: Do not reject H_0 The data do not suggest that the diet is effective.

12.7 (a)

Sample data x	Difference $D = x - 10$	Magnitude $\lvert D \rvert$	Signed rank
18	8	8	6
12	2	2	1
4	−6	6	−4
1	−9	9	−7
15	5	5	3
17	7	7	5
26	16	16	9
0	−10	10	−8
14	4	4	2

(b) H_0: Md = 10; H_a: Md > 10 α = .05   Observed value: $W^- = 4 + 7 + 8 = 19$
Critical value: $c = 8$ Decision: Do not reject H_0

12.9 (a)

Sample data x	Difference $D = x - 50$	Magnitude $\lvert D \rvert$	Signed rank
42	−8	8	−4
39	−11	11	−5
37	−13	13	−6
55	5	5	3
46	−4	4	−2
52	2	2	1
33	−17	17	−7

(b) H_0: Md $= 50$; H_a: Md < 50 $\alpha = .05$ Observed value: $W^+ = 3 + 1 = 4$
Critical value: $c = 4$ Decision: Reject H_0

12.11 H_0: Md $= 10$; H_a: Md > 10 $\alpha = .01$

Data (x)	10.3	9.2	10.6	11.1	9.5	12.0	12.2	9.1	10.7
$D = x - 10$	.3	−.8	.6	1.1	−.5	2	2.2	−.9	.7
Signed rank	3	−7	5	10	−4	12	14	−8	6

Data (x)	13.1	12.1	10.1	10.2	11.3	11.0
$D = x - 10$	3.1	2.1	.1	.2	1.3	1
Signed rank	15	13	1	2	11	9

Observed value: $W^- = 7 + 4 + 8 = 19$ Critical value: $c = 20$ Decision: Reject H_0
The data suggest that the lotion does not perform as advertised.

12.13 H_0: Md $= 520$; H_a: Md > 520 $\alpha = .01$

Data (x)	535	560	500	550	560	540	610	490	545	550	510
$D = x - 520$	15	40	−20	30	40	20	90	−30	25	30	−10
Signed rank	3	11.5	−4.5	8	11.5	4.5	16	−8	6	8	−2

Data (x)	620	460	575	485	630	600	515	640	650
$D = x - 520$	100	−60	55	−35	110	80	−5	120	130
Signed rank	17	−14	13	−10	18	15	−1	19	20

Observed value: $W^- = 4.5 + 8 + 2 + 14 + 10 + 1 = 39.5$ Critical value: $c = 43$
Decision: Reject H_0 The data suggest that the advertising campaign was successful.

12.15 Let $D = x_1 - x_2$ be expressed in tenths. H_0: Md$_D = 0$; H_a: Md$_D < 0$ $\alpha = .05$
(a)

D	1	−2	−7	4	5	−3	−14	10	13	11	−2	8	−16
Signed rank	1.5	−4	−11	8	9	−6.5	−19	15.5	18	17	−4	12.5	−21

D	−19	−3	10	−23	−15	2	−21	−8	−9	−18	−17	−1	−6
Signed rank	−24	−6.5	15.5	−26	−20	4	−25	−12.5	−14	−23	−22	−1.5	−10

Observed value: $W^+ = 1.5 + 8 + 9 + 15.5 + 18 + 17 + 12.5 + 15.5 + 4 = 101$
Critical value: $c = 110$ Decision: Reject H_0

(b) $\mu_W = n(n+1)/4 = (26)(27)/4 = 175.5$

$$\sigma_W = \sqrt{\frac{n(n+1)(2n+1)}{24}} = \sqrt{\frac{(26)(27)(53)}{24}} = 39.37$$

Observed value: $z = (101 - 175.5)/39.37 = -1.89$ Critical value: $z = -1.65$
Decision: Reject H_0
The data suggest that judge A tended to give lower performance ratings than judge B.

12.17 (a)

x	Rank (x)	y	Rank (y)
43	1	55	4
47	2	62	6
51	3	68	8
59	5	75	10
65	7		
71	9		
Sums:	27		28

(b) H_0: $Md_x = Md_y$; H_a: $Md_x \neq Md_y$ $\alpha = .10$

$U_x = 27 - (6)(7)/2 = 6$ $U_y = 28 - (4)(5)/2 = 18$ Observed value: $U = U_x = 6$

Critical value: $c = 3$ Decision: Do not reject H_0

12.19 H_0: $Md_x = Md_y$; H_a: $Md_x < Md_y$ $\alpha = .05$

Data	13.6	13.9	14.5	14.7	15.2	15.3	16.1	17.2	17.3	17.5	17.9	18.1	18.4	19.2	20.5
	x	x	y	x	x	x	y	x	y	y	y	x	y	y	y
Ranks	1	2	3	4	5	6	7	8	9	10	11	12	13	14	15

$S_x = 1 + 2 + 4 + 5 + 6 + 8 + 12 = 38$ Observed value: $U = U_x = 38 - (7)(8)/2 = 10$

Critical value: $c = 13$ Decision: Reject H_0

The data suggest that the median endurance level of soccer players is larger than that for football players.

12.21 H_0: $Md_x = Md_y$; H_a: $Md_x > Md_y$ $\alpha = .05$

Data	34.7	34.9	35.8	36.2	36.3	36.5	36.6	36.8	36.9	37.1	37.3	37.5
	y	y	y	x	y	x	y	y	x	y	x	y
Ranks	1	2	3	4	5	6	7	8	9	10	11	12

Data	37.6	37.7	37.9	38.0	38.2	38.3	38.5	38.8
	x	x	x	y	y	x	x	x
Ranks	13	14	15	16	17	18	19	20

$S_y = 1 + 2 + 3 + 5 + 7 + 8 + 10 + 12 + 16 + 17 = 81$

Observed value: $U = U_y = 81 - (10)(11)/2 = 26$ Critical value: $c = 27$

Decision: Reject H_0 The data support the spokesman's claim.

12.23 The number of runs R are obtained as follows:
(a) AAAAA BBBB $R = 2$
(b) A B A B A B A B A B A B $R = 12$
(c) AAA BBB AAA BBBB $R = 4$
(d) B AA BBB A BB A B A $R = 8$
(e) BB A BB A B AAAA BBB $R = 7$
(f) AA BB AA BB AA BB A B A $R = 9$

	c_1	c_2	R	Decision		c_1	c_2	R	Decision
(a)	2	9	2	Reject H_0	(b)	3	11	12	Reject H_0
(c)	3	12	4	Do not reject H_0	(d)	3	11	8	Do not reject H_0
(e)	3	12	7	Do not reject H_0	(f)	4	13	9	Do not reject H_0

12.25 H_0: The process is random; H_a: The process is not random. $\alpha = .05$
There are 3 runs: NNNNNN DDD NNNNN Observed value: $R = 3$
Critical values: $c_1 = 2$, $c_2 = 8$ ($n_1 = 3$, $n_2 = 12$) Decision: Do not reject H_0
There is insufficient evidence to conclude that the process is not random.

12.27 H_0: Md = 625; H_a: Md > 625 $\alpha = .05$

(a)   Let x be the number of data values larger than 625.
      Observed value: $x = 13$ Critical region: {14, 15, 16, 17, 18, 19, 20} $n = 20$
      $\alpha = .037 + .015 + .005 + .001 = .058$ Decision: Do not reject H_0

(b)

Data (x)	635	660	670	615	620	660	600	610	700	675
$D = x - 625$	10	35	45	−10	−5	35	−25	−15	75	50
Signed rank	3	13.5	16.5	−3	−1	13.5	−10	−5.5	20	18

Data (x)	615	605	650	600	665	670	645	640	655	680
$D = x - 625$	−10	−20	25	−25	40	45	20	15	30	55
Signed rank	−3	−7.5	10	−10	15	16.5	7.5	5.5	12	19

Observed value: $W^- = 3 + 1 + 10 + 5.5 + 3 + 7.5 + 10 = 40$
Critical value: $c = 60$ Decision: Reject H_0

(c)   The Wilcoxon signed–rank test is a more sensitive test than the sign test.

12.29 H_0: Md = 200; H_a: Md < 200 $\alpha = .05$

(a)   Let x be the number of data values larger than 200.   Observed value: $x = 2$
      Critical region: {0, 1} $n = 8$ $\alpha = .004 + .031 = .035$ Decision: Do not reject H_0

(b)

Data (x)	193	203	196	192	187	206	190	195
$D = x - 200$	−7	3	−4	−8	−13	6	−10	−5
Signed rank	−5	1	−2	−6	−8	4	−7	−3

Observed value: $W^+ = 1 + 4 = 5$ Critical value: $c = 6$ Decision: Reject H_0

(c)   The Wilcoxon signed–rank test is a more sensitive test than the sign test.

12.31 H_0: $Md_x = Md_y$; H_a: $Md_x < Md_y$ $\alpha = .05$

Data	40	45	55	57	65	70	71	72	74	76	77	78	82	84	88	92
	x	x	y	x	x	x	x	x	x	y	y	y	y	x	y	y
Ranks	1	2	3	4	5	6	7	8	9	10	11	12	13	14	15	16

$S_x = 1 + 2 + 4 + 5 + 6 + 7 + 8 + 9 + 14 = 56$
Observed value: $U = U_x = 56 - (9)(10)/2 = 11$
Critical value: $c = 15$ Decision: Reject H_0
The data suggest that girls tend to read with more comprehension than boys upon entering the fourth grade.

12.33 Let x and y represents resorts A and B respectively.
H_0: $Md_x = Md_y$; H_a: $Md_x < Md_y$ $\alpha = .01$

Data	30.9	32.5	35.0	36.0	39.5	40.9	42.5	44.3	47.9	48.9	53.9	61.0
	x	x	x	y	x	y	y	x	y	x	y	y
Ranks	1	2	3	4	5	6	7	8	9	10	11	12

$S_x = 1 + 2 + 3 + 5 + 8 + 10 = 29$ Observed value: $U = U_x = 29 - (6)(7)/2 = 8$
Critical value: $c = 3$ Decision: Do not reject H_0
The data do not support the claim that the median price of building lots was less in resort area A than building lots in resort area B.

12.35 H_0: The process is random; H_a: The process is not random. $\qquad \alpha = .05$

(a) There are 8 runs: LL WWW LLLLL W LLL WWWWWW LL WWW
Observed value: $R = 8$ Critical values: $c_1 = 8$, $c_2 = 19$ ($n_1 = 12$, $n_2 = 13$)
Decision: Reject H_0

(b) $\mu_R = \dfrac{2(12)(13)}{12+13} + 1 = 13.48.$ $\sigma_R = \sqrt{\dfrac{2(12)(13)[2(12)(13)-12-13]}{(12+13)^2(12+13-1)}} = 2.44$

Observed value: $z = (8 - 13.48)/2.44 = -2.25$ $\qquad$ Critical values: $z = \pm 1.96$
Decision: Reject H_0 $\qquad$ The data suggest a lack of randomness.